YOUNG MEN AND FIRE

Young Men and Fire

TWENTY-FIFTH ANNIVERSARY EDITION

Norman Maclean

Foreword by Timothy Egan

THE UNIVERSITY OF CHICAGO PRESS *Chicago & London*

The University of Chicago Press, Chicago 60637
The University of Chicago Press, Ltd., London
© 1992 by The University of Chicago
Foreword © 2017 by Timothy Egan
Published 2017
Printed in the United States of America

26 25 24 23 22 21 20 19 18 17   1 2 3 4 5

ISBN-13: 978-0-226-47545-5 (cloth)
ISBN-13: 978-0-226-45035-3 (paper)
ISBN-13: 978-0-226-45049-0 (e-book)
DOI: 10.7208/chicago/9780226450490.001.0001

Library of Congress Cataloging-in-Publication Data
Names: Maclean, Norman, 1902–1990, author. | Egan, Timothy, writer
     of foreword.
Title: Young men and fire / Norman Maclean; foreword by Timothy Egan.
Description: Twenty-fifth anniversary edition. | Chicago: The University
     of Chicago Press, 2017.
Identifiers: LCCN 2016053338| ISBN 9780226475455 (cloth:
     alk. paper) | ISBN 9780226450353 (pbk.: alk. paper) | ISBN
     9780226450490 (e-book)
Subjects: LCSH: Forest fires—Montana—Mann Gulch—Prevention
     and control. | Smokejumpers—United States. | United States. Forest
     Service—Officials and employees. | Dodge, Wag, –1955.
Classification: LCC SD421.32.M9 M33 2017 | DDC 363.37/9—dc23 LC
     record available at https://lccn.loc.gov/2016053338

♾ This paper meets the requirements of ANSI/NISO Z39.48-1992
(Permanence of Paper).

CONTENTS

As I get considerably beyond the biblical
allotment of three score years and ten, I feel
with increasing intensity that I can
express my gratitude for still being around on
the oxygen-side of the earth's crust only by
not standing pat on what I have hitherto
known and loved. While the oxygen lasts, there
are still new things to love, especially
if compassion is a form of love.

NORMAN MACLEAN
Notes written as a possible epigraph to
Young Men and Fire, *December 4, 1985*

FOREWORD

Timothy Egan

For nearly a decade, I traveled about fifty thousand miles a year roaming over the American West as a prospector of stories, on behalf of the *New York Times*. I'm a native of that oversize land, third generation, so the chance to learn from and explain the great swath of country on the sunset side of the Mississippi was a pinch-me privilege. I loved most of it: the snowstorms and the wildfires, the political squabbling and the oddball life stories, everything but the tragedies. On deadline, with a clock ticking to a pre–Internet age cutoff of 5:00 P.M. mountain time, I sometimes found myself stuck while trying to bang out eight hundred words of serviceable prose. I couldn't force the right sentences, the rhythm was off, the words felt inauthentic.

In desperation, I would reach for my talisman: a copy of *A River Runs through It and Other Stories*. This slim volume, yellowed by the sun from trips to the Southwest and frayed from panicky page plucking, was my traveling companion; I never left home without it. I would open the book to a random passage and take in a swath of Norman Maclean's gin-clear prose. It was beauty and perfection, shorn of artifice. After reading for just a few minutes, I was unstuck. From him, by example, I picked up this admonition: just try to write something clean and well. The lifelong teacher was still helping students of the written word, long after he'd retired from the University of Chicago.

"You take the way it comes to you first, with adjectives and adverbs, and cut out all the crap," he told one interviewer.

By the time his second book, *Young Men and Fire*, was published in 1992, the cult of Maclean had grown well beyond Westerners with a passion for fly-fishing and uncluttered language. He was admired for his personal story, a man who didn't take up writing until he'd reached his "biblical allotment" of time—"three score years plus ten," as he reminded people. Maclean was seventy-three when *River* was released in 1976. It is a masterpiece, to use a word he would surely dismiss as hyperbolic. Here was a professor of Shakespeare trying to come to terms with the central tragedy of his family in Montana: the murder of his younger brother, Paul, who had been beaten to death. The title story's opening sentence ("In our family, there was no clear line between religion and fly fishing") and its closing lament ("I am haunted by waters") are among the best bookends in American fiction.

"It is the truest story I ever read," noted Pete Dexter, the National Book Award–winning novelist, in a profile of Maclean in *Esquire*; "it might even be the best." Robert Redford made a film, delicately faithful to the story, starring Brad Pitt.

How could Maclean top that?

I liked this old man for the Western chip on his shoulder and for his Scottish stubbornness. "Painted on one side of our Sunday school wall were the words, God Is Love," he writes in *River*. "We always assumed that these three words were spoken directly to the four of us in our family and had no reference to the world outside, which my brother and I soon discovered was full of bastards, the number increasing rapidly the farther one gets from Missoula, Montana." He was a man of two worlds, two homes: an academic one in

Chicago, and that of the Big Sky over the cabin he had helped his father build on Seeley Lake in Montana. The West of his early life gave him all the material he needed for his later life. But because he was "Western" (a pejorative in some circles), the New York publishers turned up their collective noses at his first book. Maclean liked to say that one editor objected, "These stories have trees in them."

Oh, but he got his revenge. Struggling writers, which is to say most writers, love the story that Maclean told of getting back at one of the publishers, Alfred A. Knopf, that had rejected him earlier. Upon receiving a gushing offer to print his second book, Maclean responded with this note in 1981: "If the situation ever arose when Alfred A. Knopf was the only publishing house remaining in the world and I was the sole surviving author, that would mark the end of the world of books."

He was then about five years into trying to put together his second book, the true story of the thirteen Forest Service firefighters, all but one of them Smokejumpers, who had lost their lives in the Mann Gulch fire of August 5, 1949. Maclean would labor on it for the rest of his life. He never saw its publication. When he died in 1990, at the age of eighty-seven, *Young Men and Fire* was complete in all its parts but unfinished. Maclean had struggled for more than a decade to make sense of what happened on a hot summer afternoon in the Gates of the Mountains wilderness, off the Missouri River in Montana.

In his obituary, the *New York Times* called Maclean "a professor who wrote about fly-fishing." My employer was only partially right. For with *Young Men and Fire*, published two years after his death, Maclean succeeds in saying something true and lasting about wildfire, something true and lasting about youth, and something true and lasting about death—

his own, which fast approached, and those of the boys who fell to flame in Mann Gulch. He was, in fact, a professor who wrote about tragedy and art, and how one shapes the other.

He had been a firefighter himself; he knew his way around a Pulaski—one side of the tool an ax blade, the other a hoe, used for digging fire lines—and he could read the afternoon winds that might fuel a blowup. He thought for a long time that he would have a career in the Forest Service. Maclean was camping with his family on an island in the Bitterroot River when the largest single wildfire in American history, the Big Burn of 1910, swept over western Montana, torching three million acres in two days' time and killing nearly one hundred people. Maclean was seven. The Forest Service was five years old. The Big Burn became the agency's creation myth, sanctifying many of the young Yale School of Forestry graduates who filled the Forest Service's ranks. The fire also became a cautionary tale that would guide rangers for most of the twentieth century.

"It was frightening, as what seemed to be great flakes of white snow were swirling to the ground in the heat and darkness of high noon," writes Maclean. Thereafter, he says, the Forest Service had "1910-on-the-brain."

After working summers for the Forest Service in his teens, Maclean went to Dartmouth, where Robert Frost was one of his teachers. Maclean studied literature, though he continued to flirt, well into his twenties, with the idea of working in the woods. He started teaching full-time at the University of Chicago in 1930, acquiring his PhD in 1940. He taught Shakespeare and the British Romantic poets. He was beloved, twice earning the university's highest teaching award. The school year was spent in Chicago, summers at Seeley Lake.

It was during one of those summers, in 1949, that the

Mann Gulch fire broke out. The blaze was nothing special, even in a hot year: a couple of hundred acres burning in nearly inaccessible brush on the eastern side of the Continental Divide, twenty-seven miles north of Helena. The Smokejumpers, the elite corps that was not yet ten years old, were called on to snuff the fire before it could get any bigger. The ranks of the Smokejumpers, as Maclean noted, included many young men who fought fire in the summer and pursued their master's degrees or doctorates during the rest of the year. As a kid who spent summers in the mountains of northern Idaho and western Montana, I worshipped these guys—tough, smart, brave boys who leaped from airplanes into a vertical slope full of smoke while carrying nearly a hundred pounds of gear. They were a quick-strike force, their mission to get at the kind of gnarly fire that mere earthbound mortals could not.

But on August 5, 1949, the men from Missoula proved all too mortal. Less than two hours after leaping from a C-47 that banked low in the white summer sky, all but three of a crew of fifteen Smokejumpers were dead or fatally burned. Stoked by crosswinds, the fire in Mann Gulch blew up and leaped over onto the other side of the ravine. There, the men were trapped. They raced uphill, seeking the safety of the ridgeline. The fire roared just behind them, until it overwhelmed the crew, suffocating most of them before their bodies burned. The foreman, R. Wagner Dodge, lit an escape fire, premised on the idea that if he could create a little section of burned-over land, the big blaze at their backs would skip over that patch. Dodge ordered his men to take refuge in the burned-over area. Either they didn't hear him or didn't listen. Dodge survived, as did two men who made it over the ridge.

A few days after the blowup, which burned 4,500 acres, Maclean himself visited the site. In "Black Ghost," a story found after his death, he recalls his first trip to Mann Gulch. The fire was then—and remains to this day—the worst tragedy in the history of the Smokejumpers. What stood out during Maclean's walk over the blackened hillside was a badly burned deer, "hairless and purple," as Maclean writes. "Where the skin had broken, the flesh was in patches."

Maclean returned to the university. The Forest Service conducted a hasty review, concluding that the agency itself was not to blame for the deaths. Lawsuits were filed by family members of the dead, asserting the opposite—that Dodge's escape fire had contributed to and perhaps caused the deaths. Hollywood made a film of the burn, *Red Skies of Montana*, substituting another fire's happy ending for the tragic dissonance left over by this one. For the most part, that was that.

But Maclean could never let it go. In retirement, after the success of *A River Runs through It*, he began an earnest effort to put into print something definitive about the Mann Gulch fire. It consumed the final thirteen years of his life, and perhaps the final forty-one years of his life, going back to 1949. You can imagine him at Seeley Lake, rising for coffee as the mist lifted on the water, producing his three hundred to four hundred words by noon at the little red table in the family cabin. He was alone, having lost his wife, Jessie Burns Maclean, to cancer in 1968. "When it's good," he told Pete Dexter, "I see my life coming together in paragraphs."

As a fire book—that is, a nonfiction account of the kind of blaze that haunts western forests—Maclean's recounting is a model of scrupulous narrative journalism. He never tries to overdramatize or hype the story. It is not told from any one person's point of view, though Maclean is clearly empathetic to Dodge. He lays it out as a tale of an ordinary day in the life of western firefighting turned extraordinary in all the wrong ways. A lightning strike. A combustible punch. *Boom. Boom.*

Fire. A lookout calls it in. The Smokejumpers take to the air. At stake is not so much Mann Gulch but the stunning scenery of an area one canyon away, where the Lewis and Clark expedition had spent some time. It has value to tourists and lovers of the outdoors, and therefore must be saved from fire. The mission was to contain the blaze in the gulch.

Maclean is obsessed with getting it right—the details of smoke and wind, the ferocity of the flames, how a wildfire can create its own weather system. He is a man of the mind, an intellectual in his bristly way, who also respects those who work with their hands. His attention to detail was one of the things he was most proud of with his first book. *A River Runs through It* is fiction—the first original fiction ever published by the University of Chicago Press—informed by the real life of Maclean and his family: the roguish, beautiful brother, a master fisherman, and their father, a Presbyterian minister.

"There's no bastards in the world who like to argue more than fishermen, and not one of them corrected me on anything," Maclean said in *Esquire*. "That is my idea of a good review."

So he spends the first part of *Young Men and Fire* stating the case, following a timeline. The book is constructed as a triptych. The facts are simple and will lead Maclean to a resolution. But the facts turn out to be muddied, and a conclusion is slow to come. He goes to the Smokejumper base in Missoula, studies the science of wind and fire, pores over all documents. He discovers, somewhat early on, that the Forest Service tried to cover up some of the details—no surprise, given the ways of bureaucratic self-protection. He narrows his inquiry to what happened between 5:00 and 6:00 P.M. on August 5, in hundred-degree heat. The foreman, Dodge, is long dead from cancer a mere five years after the fire. The only two survivors are—where? Nobody knows, Maclean is told. But he tracks them down and talks them into joining him for a return to Mann Gulch in 1978. On that haunted slope, crosses mark where the men fell. From the survivors, Maclean does not get much. They were young and scared, and they ran and hid. They knew little of Dodge's fire or the fate of the others.

The next year, Maclean returns again to Mann Gulch. Now he's seventy-six years old, but he can still scramble up a Rocky Mountain incline in ninety-seven degree heat. He ends the first part of his book with several questions unresolved. Could Dodge's escape fire have saved the Smokejumpers, if they'd listened to him and taken refuge inside the burned area? Or did it burn the men themselves? And by now, the reader is wondering about Maclean himself. What's driving this old man to find these answers? He's obsessed. Why? "They were young and did not leave much behind them and need someone to remember them." But there is something more than that.

Beginning with the second part, the book becomes less of a traditional fire book (if there is such a thing) and more a forward-moving meditation, propelled by Maclean's ceaseless questioning. "Far back in the impulses to find this story is a storyteller's belief that at times life takes on the shape of art," he writes.

Helped by a former Smokejumper, Laird Robinson, who becomes his research partner, Maclean eventually arrives at a conclusion. The truth of the fire is one thing, and he feels a certain satisfaction in getting there. "If now the dead of this fire should awaken and I should be stopped beside a cross, I would no longer be nervous if asked the first and last question of life, How did it happen?"

But the truth of mortality, why and when it strikes, is

another question, one that eludes Maclean to the end and drives the literary power of *Young Men and Fire*. He's trying to shape, or at least to see, art in tragedy, while acknowledging that "tragedy is the most demanding of all literary forms." Maclean wants to give us some answers about the death of twelve men, about the loss of his wife, about his own end, in much the same way that he was trying to make something of his brother Paul's killing.

"I have been trained all my life to start by trying to make sense out of dying," he writes in 1980, in a letter to a friend. "I've come to a place in my story where it doesn't make sense anymore and (with a cold) I'm having a hard time thinking of what to write." Here, we see his doubts, which deepen with every passing month. Later on that year, he tells another friend that he won't be able to live with himself unless he can finish the book. The story is not eluding him. But the larger meaning is, he confesses. "It is clear to me now that the universe in its truculence doesn't permit itself to be that well known."

Maclean would plow on, sifting and sorting, matching some of the new science of fire mechanics with his gut instinct and his experiences in the woods. By 1984, he tells another interviewer, age has finally caught up with him. "I'm now getting so old that I can't write much anymore." With the modesty that was a trademark of his generation, Maclean undersold himself. Yes, his powers to commit words to page were receding, as would those of anyone living through his ninth decade. But a close reading of *Young Men and Fire* cannot conclude that Maclean failed to grasp what it means to die young and unfulfilled. It's there in the very pursuit of those answers—the search as the solution.

Since the Mann Gulch fire, there have been other disasters involving flame and young men in the woods of the West.

The 1994 South Canyon fire, not far from Glenwood Springs, Colorado, killed fourteen people. All the familiar elements were in place: a hot afternoon, a huff of sudden wind, a dash for safety, confusion. I waited at a base camp in Colorado to hear some news over the radio—maybe a miracle. Instead, all we got were questions—the why, why, why when people are taken at such a young age. The bodies, some of them, were found inside foiled shelter pup tents deployed as a last resort. The firefighting community, as after Mann Gulch, vowed that this kind of sacrifice would never happen again. They would learn from the loss.

And then, almost twenty years later, it happened again. The setting for the 2013 Yarnell Hill fire was different from Colorado or Montana—this fire kindled in the brush and scrub trees of Arizona—but the circumstances were not. Once again, the wildland firefighting elite, the Hotshots in this case, from another federal agency, were trapped by shifting winds and a blowup. Communication was muddled. With no escape route, nineteen members of the Granite Mountain Hotshots were killed—the greatest loss of life for a federal crew since the Big Burn of 1910.

Maclean would surely shake his head at these tragedies, the lessons unlearned, the history repeating itself, the post-fire motions of grief, exasperation, and denial. No matter how meticulously he detailed the mistakes that led to Mann Gulch, he surely knew that loss of life is always a possibility when human beings are thrown at flame. We are predictable; wildfire is not. You can read this book as a great warning and find much to incorporate into the evolving wisdom of the firefighting canon. In that regard, it will live for many years. But you can also read this book as a meditation on the inevitable fate of us all. And in that regard, Maclean has given us what the best writers offer: the immortality of his words.

PUBLISHER'S NOTE

Though he had hoped for many years to write about the Mann Gulch fire, Norman Maclean did not start work on this book until his seventy-fourth year, after publication of *A River Runs through It and Other Stories*. He began *Young Men and Fire* partly in the spirit of what he liked to call his "anti-shuffleboard" philosophy of old age, but partly, too, out of a deeper compulsion. In Maclean's files after his death were found some notes toward a preface, written in 1984. "The problem of self-identity," Maclean wrote, "is not just a problem for the young. It is a problem all the time. Perhaps the problem. It should haunt old age, and when it no longer does it should tell you that you are dead." *Young Men and Fire* was where, near the end, all the lives he had lived would merge: the lives of a woodsman, firefighter, scholar, teacher, and storyteller.

When Maclean died in 1990 at the age of eighty-seven, *Young Men and Fire* was unfinished. The book had resisted completion because the facts of the catastrophe proved so protean and because Maclean's stamina began finally to wane. But more important, *Young Men and Fire* had become a story in search of itself as a story, following where Maclean's compassion led it. As long as the manuscript sustained itself and its author in this process of discovery, it had to remain in some sense unfinished.

After Maclean's death, it was left for the Press to prepare *Young Men and Fire* for publication. Our editing has not altered the structure of the book, and we have kept substantive interpolations to a minimum. We have done the kind of stylistic editing that we believe Maclean himself would have done if he had had the time, and we have cut certain repetitions in the manuscript. Facts have been checked for consistency and accuracy and occasionally corrected, but they have not been updated beyond 1987, the year Maclean became too ill to work further on the manuscript. We have added the present chapter divisions, although the breaks within these chapters are Maclean's, as is the division of the book into three parts. "Black Ghost," the story that opens this book, was found in Maclean's files after his death, his exact intentions for it unclear. We print it here as a fitting prelude.

Norman Maclean talked much of the loneliness of writing, but he also relished what he called its social side, and he planned to acknowledge the help he had received in writing this story. His greatest debts are recorded in the story itself: to Laird Robinson, Bud Moore, Ed Heilman, Richard Rothermel, Frank Albini, and other men of the United States Forest Service; to women of the Forest Service, among them Susan Yonts, Beverly Ayers, and Joyce Haley; and to the survivors of the Mann Gulch fire, Walter Rumsey and Robert Sallee. Maclean would have thanked dozens more.

In editing the manuscript, the Press has benefited from the advice, at various stages, of Wayne C. Booth, Jean Maclean Snyder, and John N. Maclean. Laird Robinson was Norman Maclean's partner in his quest for the missing parts of the Mann Gulch story, and we thank him for helping the Press in the same spirit. We are grateful, too, for the assistance of Joel Snyder, Dorothy Pesch, William Kittredge, Wayne Williams, and Richard Rothermel. We especially thank Jean Maclean Snyder and John N. Maclean for entrusting *Young Men and Fire* to the University of Chicago Press and working with us to bring it to publication.

YOUNG MEN AND FIRE

Black Ghost

It was a few days after the tenth of August, 1949, when I first saw the Mann Gulch fire and started to become, even then in part consciously, a small part of its story. I had just arrived from the East to spend several weeks in my cabin at Seeley Lake, Montana. The postmistress in the small town at the lower end of the lake told me about the fire and how thirteen Forest Service Smokejumpers had been burned to death on the fifth of August trying to get to the top of a ridge ahead of a blowup in tall, dead grass. In the small town at Seeley Lake and in the big country around it there are only summer tourists and loggers, and, since the loggers are the only permanent residents, they have all the mailboxes at the post office—the postmistress, of course, has come to know them all, and as a result knows a lot about forests and forest fires in a gossipy way. Since she and I are old friends, I have a box, too, and every day when I came for my mail she passed on to me the latest she had heard about the dead Smokejumpers, most of them college boys, until after about a week I realized I would have to see the Mann Gulch fire myself while some of it was still burning.

I knew, of course, that a fire that big would be burning long after it had been brought under control. I had gone to work for the Forest Service during World War I when there was a shortage of men and I was only fifteen, four years younger than Thol, the youngest of those who had died in Mann Gulch, so by the time I was his age I had been on several big fires. I knew, for instance, that the Mann Gulch fire would be burning for a long time, because one November I had gone back with my father to hunt deer in country close to where I had been on a big fire that summer, and to my surprise I had seen stumps and fallen trees still burning, with smoke coming out of blackened holes in the snow.

But even though I knew smoke would probably be curling out of Mann Gulch till November there came a day in early August when I could not listen to any more post office gossip about the fire. I even had a notion of why I had to go and see the fire right then. I once had seen a ghost, and the ghost again possessed me.

The big fire that had still been burning late into the hunting season had been on Fish Creek, the Fish Creek that is about nine miles by trail, as I remember, from Lolo Hot Springs. Fish Creek was fine deer country, and the few homesteaders who had holed up there made a living by supplementing the emaciated produce from their rocky gardens with the cash they collected from deer hunters in the autumn by turning their cabins into overnight hunting lodges. Deer, then, were a necessary part of their economy and their diet. They had venison on the table twelve months a year, the game wardens never bothering them for shooting deer out of season, just as long as they didn't go around bragging that they were getting away with beating the law.

Those of us on the fire crew that had been sent from the ranger station at Lolo Hot Springs were pretty sure that the fire had been started by one of these homesteaders. The Forest Service had issued a permit to a big sheep outfit to graze a flock of a thousand or so on a main tributary of Fish Creek, and you probably know—hunters are sure they know—that sheep graze a range so close to the ground that nothing is left for a deer to eat when the sheep have finished. Hunters even say that a grasshopper can't live on the grass sheep leave behind. The fire had been started near the mouth of the tributary, on the assumption, we assumed, that the fire would burn up the tributary, which was a box canyon, all cliffs, with no way of getting sheep out of it. From a deer hunter's point of view, it was a good place for sheep to die. The fire, though, burned not only up the tributary but down it to where it entered Fish Creek and could do major damage to the country. We tried first to use Fish Creek as a "fireline," hoping to stop the fire at the water's edge, but when it reached thick brush on one side of the creek it didn't even wait to back up and take a run before it jumped into the brush on the other side. Then we were the ones who had to back up fast. At this point, Fish Creek is in such a narrow and twisted canyon that the main trail going down it is on the sidehill, so we backed up to the sidehill trail, which was to be our second line of defense.

I was standing where the fire jumped the trail. At first it was no bigger than a small Indian campfire, looking more like something you could move up close to and warm your hands against than something that in a few minutes could leave your remains lying in prayer with nothing on but a belt. For a moment or two I could have stepped over it and fought it just as well from the upstream side, and when it got a little bigger I still could have walked around it. Instead, I fought it where I stood, for no other reason than that all of us are taught to be the boy who stood on the burning deck. It never occurred to me that I had alternatives. I did not even notice—not until I returned the next day—that if I had stepped across the fire I would have been on a side of it where the fire would soon reach a cedar thicket whose fallen needles had made a thick, moist duff in which fire could only creep and smolder.

The fire coming up at me from the creek in the bunch and cheat grass stopped for only a moment when it reached the trail we were hoping to use as a fire-line. The grass on either side of the trail did not make such instant connections as the brush had on the sides of the creek. Here the fire rocked back and forth like a broadjumper before it started toward the takeoff. Then it jumped. One by one, other like fires reached the line, rocked back and forth, and they all made it.

I broke and started up the hillside. Unlike the boys on the Mann Gulch fire, who did not start running until they were nearly at the top, I started running near the bottom. By the testimony of those who survived, they weren't scared until the last hundred yards. My testimony is that I was scared until I got near the top, when all feelings—fright, thirst, desire to stop for a moment to pray—became indistinguishable from exhaustion. Unlike the Mann Gulch fire, though, the fire behind was never quite a blowup; it was never two hundred feet of flame in the sky. It was in front of me, though, as well as behind me, with nowhere to go but up. Above, it was little spot fires started by a sky of burning branches. The spot fires turned me in my course by leaping into each other and forming an avalanche of flame that went both down and up the mountain. I kept looking for escape openings marked by holes in smoke that at times burned upside down. Behind, where I did not dare to look, the main fire was sound and heat, a ground noise like a freight train. Where there were weak spots in the grass, it sounded as if the freight train had slowed down to cross a bridge or perhaps to enter a tunnel. It could have been doing either, because in a moment it roared again and started to catch up. It came so close it sounded as if it were cracking bones, and mine were the only bones around. Then it would enter a tunnel and I would have hope again. Whether it rumbled or crackled I was always terrified. Always thirsty. Always exhausted.

Halfway or more toward the top I heard a voice beside me

when the roar of the main fire was reduced for a moment to the rattle of empty railroad cars. The voice sidehilled until it was on my contour and said, "How're you doing, sonny?" The voice may have come down with a burning branch, or it may have belonged to a member of our pickup crew whom I had never seen before. The only thing I noticed about him at first was that he didn't slip because he wore a good pair of climbing boots with caulks in them.

In answer to his question of how sonny was doing, I answered, "I keep slipping back," pointing back but not looking back. I also pointed at my shoes.

I was in my second summer in the Forest Service, so I knew what good climbing shoes were and how hobnails in them weren't enough to make them hold on hillsides, especially on slick, grassy hillsides, but I was young and still trying to escape such harsh realities as growing up and paying half a month's wages for a good pair of shoes. Consequently, I had gone to an army surplus store and bought a pair of leftover shoes from World War I. They were cheap shoes and wouldn't hold long caulks so I had rimmed them with hobnails, and hobnails soon wear as smooth as skates. The ghost in the caulks climbed straight uphill and never slipped, but I had to weave back and forth in little switchbacks and dig in with the edges of my soles.

Being sorry for myself made me feel that I couldn't go any farther; being terrified made me feel exhausted; waiting for the ghost in caulked shoes to help me made me feel exhausted; being so thirsty that I couldn't form words to ask for help made me feel exhausted. As a fire up a hillside closes in, everything becomes a mode of exhaustion—fear, thirst, terror, a twitch in the flesh that still has a preference to live, all become simply exhaustion. So upon closer examination, burning to death on a mountainside is dying at least three times, not two times as has been said before—first, considerably ahead of the fire, you reach the verge of death in your boots and your legs; next, as you fail, you sink back in the region of strange gases and red and blue darts where there is no oxygen and here you die in your lungs; then you sink in prayer into the main fire that consumes, and if you are a Catholic about all that remains of you is your cross.

The black ghost that could walk in a straight line came closer to me and took a look. I looked back but out of fright. The black ghost had a red face. In more leisurely times he could have been an alcoholic, but certainly much of the red now was reflection from the flames. "Could I be of any help?" the red face asked, becoming a voice again.

I thought I was beyond help, but I swallowed the thirst in my throat to find a word and said, "Yes." When I was able, I said, "Yes, thanks."

The black ghost came closer, the red in his face burning steadily. Then suddenly in his face there was a blowup, a reflection of something either behind him or in him, and he slapped me in the face.

"My God," I said, and reeled sideways across the hillside. All I knew while I staggered was that if I fell I might never get up. I burst into tears even while I was staggering, but came to rest standing. When I could recover my breath and hold the hot air in my lungs, I cried out loud, especially when I realized that all that could save me now was my army surplus boots. Still in tears, I proceeded to climb almost straight uphill almost without slipping until I reached the contour where I had been outraged. Here I stopped and went looking for what had done it, but he was far up the hill, peering down at me from the mouth of a cave that now and then opened in a red cliff of flame.

This is all I know about the violent apparition that was

just ahead of the fire. It must have been a member of the fire crew that had been picked up in the bars of Butte. I had never seen him before and have never seen him since. Maybe he was a Butte wino demented by thirst. Maybe he was a Butte miner with tuberculosis whose lungs were collapsing wall to wall from the heat of the air they were trying to breathe. It is even possible something was working on him besides the fire as he tried to keep ahead of it, something terrible that had been done to him for which he had to get even before his consignment to flames. In either of these cases, at just the right moment I may have appeared out of smoke, young and paralyzed and unable to do anything back if he did what he wanted to do, so he did it.

Above, the crevice in the red cliff opened now and then. From its entrance a figure retreated and ascended into the sky until he hung like a bat on the roof of a cave. He was always watching me, but I don't know what he hoped to see. Finally, he was hanging upside down by his claws.

I have no clear memory of going up the rest of the ridge, except that when I reached the top I had to put out the fire that smoldered in my shoelaces. I didn't think of the crew or where I might find them. I didn't think of the ghost. After I reached the top of the ridge, for a time I couldn't think of anything behind me. I thought of things ahead, the nearest of which was a hunting lodge up the main fork of the Fish Creek where my father and I had stayed the last two hunting seasons. It was run by a woman, Mrs. Brown, who looked something like my mother but more like an Indian with brown crow's-feet in the corners of her eyes. In addition, she was a fine shot so that if one of her guests didn't get a deer she would run out for an hour or two the last morning and shoot one for him. I thought, if I follow the ridge upstream and then drop into the creek bottom where her cabin is, Mrs.

Brown will be able to do something for me, even if I am in pretty bad shape. I thought she might even shoot a deer for me, and then I thought, no, that's wrong—I don't need a deer, but go anyway. You need something.

Now that I had started thinking again I became exhausted again. Mrs. Brown can help men who cannot help themselves, I thought, and I am exhausted beyond comprehension. It took me until nearly dark to get within a quarter of a mile from her cabin. Then I put exhaustion out of mind and ran the last quarter of a mile to get there in time, although time was just a hangover from the past with no present meaning. I did not collapse, but I rocked on my feet from the suction of air she caused when she opened the cabin door.

She did not ask a question. She said quietly, "Come in and lie down. You look very white." I was baffled. I was sure I was black. "No," she said, "you're very white."

She felt the water in the pail but it was evidently lukewarm, so she went down to the creek and dipped out a cold pail. Then she said, "I told you to lie down." She still hadn't asked a question. If you have spent your life in a cabin, you know that there are times when you have to do things before you try to find out what things are all about.

She washed me in cold water again and again, taking my pulse each time she did. Then finally she took my pulse again, nodded her head, and threw a whole dipper of water on me to signify my convalescence period was over. She buttoned my shirt and said, "When you go deer hunting this autumn, you'll get your limit."

Not until then did she try to find out what had happened to me and how I had got there. "Did you try to stop the fire?"

"Mrs. Brown?" I asked. "Mrs. Brown, did you start it?" I finished asking.

"You'll have no trouble getting your limit next deer season," she reassured me.

I knew that she wasn't going to say anything more and that I'd better not ask anything more.

"For a preacher," she said, "your father is very handy with a rifle."

I said, "He's also good with a scatter-gun."

Then she said, "I'll write him tomorrow and let him know what happened."

I said to her, "Skip it. He knows I can take care of myself."

I wasn't sure I should have come to her for help, now that I didn't need help anymore. Even tough women who are good with the rifle get motherly when it's all over.

She advised me, "Stay here and take it easy until tomorrow. Then go back to Lolo Hot Springs and report to the ranger station there."

After a pail and a half of cold water I no longer needed an Indian mother, and I wanted to make this clear. I said, even if she was a good shot, "Thanks, I'll stay over tonight, but tomorrow I'll go back to the fire and see if I can find the crew." I didn't add that mostly I wanted to see where the fire had jumped the trail and I had started up the ridge ahead of it, fear being only partly something that makes us run away—at times, at least, it is something that makes us come back again and stare at what made us run away.

Although I was very tired the next morning, I hurried to where the fire had jumped the trail. It was not until I stared that I realized if I had stepped across a little fire and been on the side of it where the cedar duff was that I would not have had to race a giant uphill for my life or been stopped by a ghost along the way.

I have several times returned to that spot on the trail. The autumn of the same year of the fire I said to my father,

"Let's hunt another season at Mrs. Brown's in Fish Creek." It was while hunting in Fish Creek that November that I saw smoke coming out of black holes in the snow.

While I listened to the postmistress in the record heat of early August of 1949, my memory turned to snow with smoke curling out of it. When the smoke started mixing at night with my dreams, I locked my cabin and drove the 150 miles to Wolf Creek and picked up my brother-in-law, who had fought a few days on the Mann Gulch fire as a volunteer. He and I borrowed a Dodge Power Wagon from the Oxbow Ranch, because we knew it would be tough going ahead. Then we drove up the dirt road on the east side of the Missouri until we came to the Gates of the Mountains and to Willow Creek, the creek north of Mann Gulch where the fire was finally stopped on its downriver side. The road went up the creek for a way and then dead-ended, and it and the creek had been used by the crew—on the whole, successfully—as fire-lines. Only occasionally had the fire jumped both of them, and where it had the crew quickly controlled it again. By the time we reached the fire the crew had put it out for several hundred yards back of the road or creek, sawing down the still-burning trees and either burying them with dirt or pouring water on them when they were near the creek. Since there was no danger of the fire jumping the lines again, the crew had moved on, leaving the fire to burn itself harmlessly in its interior.

It was a world of still-warm ashes that had incubated once-hot poles. The black poles looked as if they had been born of the gray ashes as the result of some vast effort at sexual intercourse on the edges of the afterlife. When the vast effort was over, it was discovered the poles were born dead and the ashes themselves lived only because the winds moved them. It was the amphitheater of the afterlife where passion had destroyed life, but passion devoid of life could be reborn. A little farther into the fire a black pole would now and then explode and reproduce a progeny of flames. A cliff would tear loose a tree it had kept burning in a secret crevice and then toss the sacrifice upon the rocks below, where the victim exploded into flames and passion without life. On the fire-lines of hell sexual intercourse seems to be gone forever and then brutally erupts, and after a great forest fire passes by, there are warm ashes and once-hot poles and passion in death.

Not far up the creek my brother-in-law stopped the Power Wagon. Ahead, the creek came close to the road, and standing in the creek was a deer terribly burned. It was drinking and probably had been for a long time. It was probably like the two Smokejumpers, Hellman and Sylvia, who did not die immediately and could never put out their thirst, drinking at every chance until they became sick at the stomach.

The deer was hairless and purple. Where the skin had broken, the flesh was in patches. For a time, the deer did not look up. It must have been especially like Joe Sylvia, who was burned so deeply that he was euphoric. However, when a tree exploded and was thrown as a victim to the foot of a nearby cliff, the deer finally raised its head and slowly saw us. Its eyes were red bulbs that illuminated long hairs around its eyelids.

Since it was August, we had not thought of taking a rifle with us, so we could not treat it as a living thing and destroy it.

While it completed the process of recognizing us, it bent down and continued drinking. Then either it finally recognized us, or became sick at the stomach again.

It tottered to the bank, steadied itself, and then bounded off euphorically. If it could have, it probably would have said, like Joe Sylvia, "I'm feeling just fine." Probably its sensory apparatus, like Joe Sylvia's, had been dumped into its bloodstream and was beginning to clog its kidneys.

Then, instead of jumping, it ran straight into the first fallen log ahead.

My brother-in-law said, loathing himself, "I forgot to throw a rifle into the cab of the truck."

The deer lay there and looked back looking for us, but, shocked by its collision with the log, it probably did not see us. It probably did not see anything—it moved its head back and forth, as if trying to remember at what angle it had last seen us. Suddenly, its eyes were like electric light bulbs burning out—with a flash, too much light burned out the filaments in the bulbs, and then the red faded slowly to black. In the fading, there came a point where the long hairs on the eyelids were no longer illuminated. Then the deer put its head down on the log it had not seen and could not jump.

In my story of the Mann Gulch fire, how I first came to Mann Gulch is part of the story.

Young Men
and Fire

PART ONE

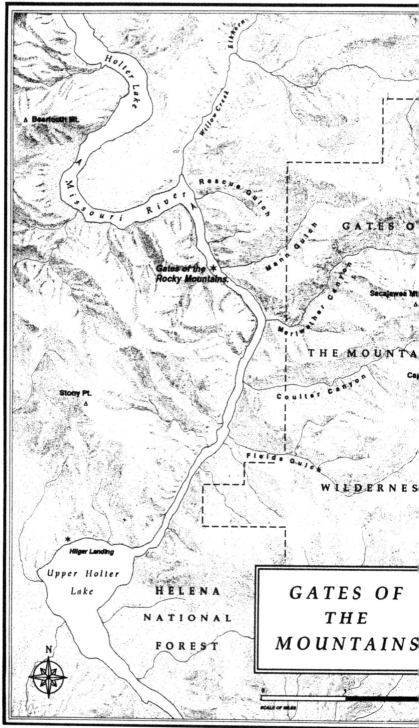

Holter Lake

△ Beartooth Mt.

Missouri River

Willow Creek

Elkhorn

Rescue Gulch

Mann Gulch

GATES O

Sacajawea Mt △

Meriwether Canyon

Gates of the
Rocky Mountains ★

THE MOUNTA

Ca

Coulter Canyon

Stony Pt. △

Fields Gulch

WILDERNES

★
Hilger Landing

Upper Holter
Lake

HELENA

NATIONAL

FOREST

N

**GATES OF
THE
MOUNTAINS**

0   5

SCALE OF MILES

TOM WILLCOCKSON

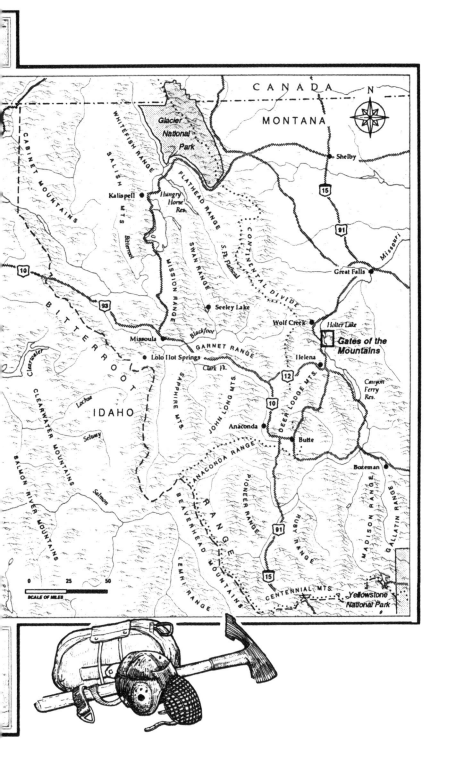

1

In 1949 the Smokejumpers were not far from their origins as parachute jumpers turned stunt performers dropping from the wings of planes at county fairs just for the hell of it plus a few dollars, less hospital expenses. By this time they were also sure they were the best firefighters in the United States Forest Service, and although by now they were very good, especially against certain kinds of fires, they should have stopped to realize that they were newcomers in this ancient business of fighting forest fires. It was 1940 when the first parachute jump on a forest fire was made and a year later that the Smokejumpers were organized, so only for nine years had there been a profession with the aim of taking on at the same time three of the four elements of the universe—air, earth, and fire—and in a simple continuous act dropping out of the sky and landing in a treetop or on the face of a cliff in order to make good their boast of digging a trench around every fire they landed on by ten o'clock the next morning. In 1949 the Smokejumpers were still so young that they referred affectionately to all fires they jumped on as "ten o'clock fires," as if they already had them under control before they jumped. They were still so young they hadn't learned to count the odds and to sense they might owe the universe a tragedy.

It is true, though, that no technical advance was to influence the Forest Service's methods of spotting and fighting wildfires as much as the airplane, which arrived early in the century about the same time as the Forest Service (1905). Two world wars hastened the union between airplanes and firefighting. By 1917 chief forester Henry S. Graves was conferring with the chief of the Army Air Corps about the possibility of army planes flying patrol missions over western forests. By 1925 the Forest Service itself started using planes from which fires could be spotted more quickly and thoroughly than from scattered lookouts. By 1929 planes were dropping supplies to firefighters, and it seemed that soon firefighters themselves would be dropped, but psychological difficulties and difficulties with equipment held back the development of parachute jumping on wildfires. It was only after several years of experimenting and training that the first parachute jump on a forest fire was made, one of the two jumpers being Earl Cooley, who was to be the spotter on the C-47 that carried the Smokejumpers to the Mann Gulch fire and, as spotter, tapped each jumper on the left calf as the signal to step into the sky over Mann Gulch.

The chief psychological roadblock holding up the acceptance of parachute jumping by the government and the public itself was the belief that most parachute jumpers were at least a little bit nuts and the high probability that a few of them were. In 1935, Evan Kelley, of the Forest Service's Region One (with headquarters in Missoula, Montana, where in a few years one of the biggest Smokejumper bases was to be established), rejected the possibility of dropping men on fires from parachutes by saying: "The best information I can get from experienced fliers is that all parachute jumpers are more or less crazy—just a little bit unbalanced, otherwise they wouldn't be engaged in such a hazardous undertaking." There is no doubt that among those most visibly touched with the Icarus complex were jumpers off wings of planes at county fairs or stuntmen doing the same kind of work for movies. Only a year before Kelley had made his psychological analysis of parachute jumpers, Frank Derry, a stuntman in

California and short of cash, got the idea of jumping from a plane in a parachute, dressed as Santa Claus. He made a perfect landing, pleased the local Los Angeles merchants, quit factory work for good, joined a flying circus barnstorming the West, and became one of the nine original Forest Service Smokejumpers, one of the Forest Service's finest jump instructors, and one of its best riggers, making important improvements in both the parachute and the jump suit.

Most people have a touch of the Icarus complex and, like Smokejumpers, wish to appear on earth from the sky. In my home town of Missoula, Montana, older brothers all over town trained their younger brothers to jump from garage roofs, using gunnysacks for parachutes. The older brothers argued that the younger brothers should do the jumping because, being smaller, they would take longer to reach the ground and so give their gunnysacks more time to open and soften the landing. From the start, Smokejumpers had to have a lot of what we have a little of, and one way all men are born equal is in being born at least a little bit crazy, some being more equal than others. A number of these latter were needed to get the Smokejumpers started, and a certain number more have always been needed to keep it going.

Fortunately, many of those powered by the Icarus complex, unlike Icarus, are gifted mechanically in odd ways and have long worked on problems connected with landing safely. Even the most sublime of oddballs, Leonardo da Vinci, had studied the problem of safely landing men on earth from the sky. But it wasn't until 1783 that the French physicist Louis-Sébastien Lenormand made the first successful parachute jump from a tower, and even in 1930 the parachute had many shortcomings as a means of aerial transport, some of which were eliminated or reduced by none other than Frank Derry, the Santa Claus parachute jumper who was also gifted mechanically. One of the parachute's greatest shortcomings as aerial transport had been that, being a parabolic object, it drops with a bell-like motion. As it descends, air is forced up into it and, since there are no openings in the parachute through which the air can escape, it rocks up on one side until the surplus air is released, then swings to the other side until it tips out the excess air it has accumulated on its return trip. As a result, before the parachute could be a reasonably safe means of getting from the sky to the earth, the rocking had to be taken out of its flight and some means of steering it had to be devised so Smokejumpers and their supplies could be dropped on a designated spot near a fire instead of scattered all over the nearby mountains.

The parachute developed by Frank Derry became the standard Smokejumpers' parachute for many years and is the parachute used by the crew that dropped on the Mann Gulch fire. The rocking motion had been reduced by three openings through which air could be released—an opening in the top and two slots on opposite sides. On the outside of the chute attached to the slots were "tails," pieces of nylon that acted as rudders to guide the flow of air coming through the slots, and to them guide lines were attached so that the direction of the flight was ultimately determined by the jumper. Not a highly safe and sensitive piece of machinery, but better than Icarus had. It had a speed of seven or eight knots, and, as soon as a jumper could, he turned his face to the wind and looked over his shoulder to see, among other things, that he didn't smash into a cliff.

Frank Derry, his two brothers, and others of the early Smokejumpers not only greatly improved the parachute but soon were developing a safer jump suit, one designed especially for jumping in mountainous timber country—football helmet with heavy wire-mesh face mask, felt-lined suit, and "shock absorbers" such as ankle braces, athletic supporters, back and abdominal braces, and heavy logger boots (the White logger boots from Spokane, Washington, the best). Frank Derry's two brothers were helpful, but staying put was not part of their calling and they weren't long with the Smokejumpers. Frank, however, lasted much longer, then bought a bar nearby and became his own best customer.

So far it has all been the jump in smokejumping and nothing about the smoke or fire at the end. In 1949 a fair number of old-timers in the Forest Service still believed that God means there to be only one honest way to get to a forest fire and that is to walk your guts out. To these old-timers the Smokejumpers were from a circus sideshow, although in fact they were already on their way to becoming the best firefighting outfit in the Forest Service.

Basic movements in the history of the Forest Service had helped put the Smokejumpers by 1949 on their way to being the best. The United States Forest Service was officially established in 1905 by President Theodore Roosevelt and Governor Gifford Pinchot of Pennsylvania, eastern outdoorsmen who knew and loved the earth in its wondrous ways when left to itself and given a chance. Their policy of acquiring and protecting some of the earth's most beautiful remaining parts became the Forest Service's primary purpose. Then came 1910, the most disastrous fire year on record. In western Montana and Idaho 3 million acres were left behind as charred trees and ashes that rose when you walked by, then blew away when you passed. This transformation occurred largely in two days, August 20 and 21, when thousands of people thought the world was coming to an end, and for eighty-seven people it did.

I remember these two days very well. My family was on summer vacation, camped in tents on an island between forks of the Bitterroot River. The elders in my father's church had become alarmed and had come in a wagon to rescue us. A team of the elders waded out to our island, crossed hands, and in this cradle carried my mother back to the wagon. My father and I followed, my father holding me with one hand and his fishing rod with the other and I also holding my fishing rod with my other hand. It was frightening, as what seemed to be great flakes of white snow were swirling to the ground in the heat and darkness of high noon. I was seven years old and might have cried for our tent, which we had to leave behind, except I thought my mother and our two rods would make it to the wagon.

Since 1910, much of the history of the Forest Service can be translated into a succession of efforts to get firefighters on fires as soon as possible—the sooner, the smaller the fire. If a campfire left burning can be caught soon enough, a man with a shovel can bury it. If the fire is a lightning fire burning in a dead tree, a man will need an ax to drop the burning tree and will still need the shovel to dig the shallow trench into which he is going to drop and bury it. If two Smokejumpers had reached the Mann Gulch fire the afternoon it started, they would at least have kept it under control until a larger crew arrived. Before the Mann Gulch fire was finally put under control five days later, there were 450 men on it and they didn't have as much to do with stopping it as did cliffs and rock slides.

So history went from trails and walking and pack mules to roads and trucks up every gulch to four-wheel drives where there weren't any roads to planes and now to helicopters, which can go about anywhere and do anything when they get there or on the way. The Smokejumpers are a large part of this history. Graphs prove it.

The two graphs reproduced in this chapter are a part of

statistical studies by Charles P. Kern, fire coordinator of the Forest Service's Region One, and assistant fire coordinator Ronald Hendrickson of the variations in number and size of forest fires in Region One from 1930 to 1975. The first graph, "Total Number of Fires per Year," shows just what an old-time woodsman who has long fought fires would expect—that there has been no significant trend either up or down in the number of wildfires during those forty-five years. There has been a bad fire year now and then, as in the late thirties and early sixties, and there probably always will be now and then, let's hope never as bad as 1910, but on a statistical curve lightning seems to be a fairly fixed feature of the universe, as does the number of people who are careless with campfires. The result is no discernible downtrend in the number of wildfires.

The graph entitled "Number of Fires Rated Class C and Larger" tells a very different story and shows clearly the coming and continuing presence of the Smokejumpers. The number of fires rated Class C (ten to ninety-nine acres) and larger in Region One, figured as a percentage of the total fires per year, plunged sharply as the Smokejumpers became an organization in the early 1940s, then made its last sharp rise to almost 9 percent with the coming of World War II when the Smokejumpers became a depleted operation, but plunged just as precipitously when the war was over and veterans filled up the crews of Smokejumpers, who again were stopping fires before they spread far. Since 1945 there has been no year when 5 percent of the fires became Class C or larger—thirty years is surely a trend, no doubt one that cannot be ascribed solely to the Smokejumpers but one that has to be a great tribute to them.

Although this trend has to be a tribute in part to the fixed theory of doing everything possible between heaven

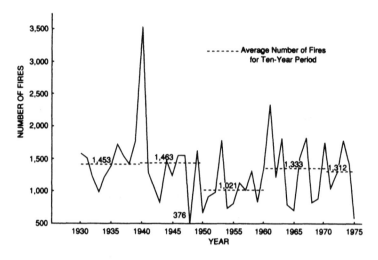

Total number of fires per year, 1930–75, in Region One of the
United States Forest Service.

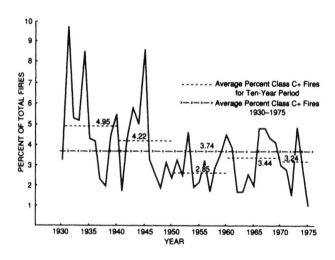

Number of fires rated Class C and larger, figured as a percentage
of total fires per year, 1930–75, in Region One of the United States
Forest Service.

and earth to get firefighters on a fire as fast as possible, what also makes a world of difference is the kind of men who get there first. The requirements used in selecting the first crews of Smokejumpers give a rough profile of the kind of men the Forest Service thought were needed to join sky with fire, and these same requirements should have given the jumpers some idea of their life expectancy. They had to be between twenty-one and twenty-five, in perfect health, not married, and holding no job in the Forest Service as important as ranger. So basically they had to be young, tough, and in one way or another from the back country. And the Forest Service carried no insurance on them.

It is not hard to imagine why the Smokejumpers from the start have had several visible bloodlines. With their two major activities—to jump from the sky and fight fire when they land—they have always drawn professional adventurers. The three Derry brothers are good examples. They were important in giving shape and substance to the early history of the Smokejumpers, and, from the nature of things, the Smokejumpers will probably always draw their quota of adventurers. On weekends, they are likely to rent a Cessna 180 and go jumping just for the hell of it; they try to make big money in the summer and some go to Honolulu and shack up for the winter, at night passing themselves off as natives to multinational female tourists or even to female natives. Others spend the winters as ski instructors in Colorado or Utah or Montana, colder work in the day but probably not at night.

One might assume that most Smokejumpers come from the woods and after they are finished as jumpers join up for good with the United States Forest Service or some state agency supervising public lands or some private logging company—the Smokejumper base in Missoula is a magnet for tough young guys pointed toward the woods for life. Besides being the headquarters for Region One of the Forest Service, Missoula is also the home of the University of Montana, which has a powerful school of forestry. Any summer a highly select number of forestry school students are Smokejumpers—of the thirteen firefighters who died in Mann Gulch, five were forestry students at the University of Montana and two were forestry students at the University of Minnesota. Two of the three survivors had just finished high school and were also University of Montana students. Select, very good students, trained in the woods.

At best, though, there is very little chance of a longtime future in smokejumping. To start with, you are through jumping at forty, and for those who think of lasting that long there are only a few openings ahead, administrative or maintenance. But one thing that remains with Smokejumpers, no matter where they ultimately land, is the sense of being highly select for life and of belonging for life to a highly select outfit, somewhat like the Marines, who know what they are talking about when they speak of themselves as the proud and the few. Although many Smokejumpers never see each other after they leave the outfit, they remain members of a kind of fraternal organization that also has some dim ties seemingly with religion. Just being a first-class woodsman admits you almost anywhere into an international fraternity of sorts, and although you will meet only a few of your worldwide brotherhood, you will recognize any one of them when you see him swing an ax. Going a little up the fraternal ladder is being admitted to the Forest Service, and that is like belonging to the Masons or the Knights of Columbus; making the next step is becoming a Smokejumper, and that is like being a Shriner or Knight Templar. This kind of talk is going too far but not altogether in the wrong direction. It is very important to a lot of people to make unmistakably clear to themselves and to the universe that they love the universe but are not intimidated by it and will not be shaken by it, no matter what it has in store. Moreover, they demand something from themselves early in life that can be taken ever after as a demonstration of this abiding feeling.

So it shouldn't be surprising that many Smokejumpers never intend to remain Smokejumpers or even to work in the woods for the rest of their lives. A good number of them are students working for M.A.'s or Ph.D.'s—even more go on to be lawyers and doctors, and even more to be dentists. These young men are first class, both as students and jumpers. They tend to hang out together but don't talk much about their university life, at least not when other jumpers are around. Later, though, when they are far away and far up the professional ladder, they get a remote look in their eyes when they talk about the tap on the calf of the left leg telling them it's only a step to the sky.

For many former Smokejumpers, then, smokejumping is not closely tied up with their way of life, but is more something that is necessary for them to pass through and not around and, once it is unmistakably done, does not have to be done again. The "it" is within, and is the need to settle some things with the universe and ourselves before taking on the "business of the world," which isn't all that special or hard but takes time. This "it" is the something special within that demands we do something special, and "it" could be within a lot of us.

On the bottom line, this is the story of an "outfit," as men call themselves when they take on the same tough job, have to be thought a little bit crazy to try it, have to stick together and share the same training to get it done, and shortly afterwards have to go to town together and stick together if one of them starts losing a fight in a bar. They back each other and they imitate each other. It should be clear that this tragedy is not a classical tragedy of a monumental individual crossing the sword of his will with the sword of destiny. It is a tragedy of a crew, its flaws and grandeurs largely those of Smokejumpers near the beginning of their history. Their collective character counts, and being young counts, it especially counts, but only certain individuals emerge out of the smoke and roar that took in everything. Eldon Diettert counts; he was the fine research student who was called from his birthday dinner to make this flight and told some of the crew that he almost said no—only recently a scholarship in the forestry school at the University of Montana has been named in his honor. David Navon was already something of a four-dimensional adventurer; he had been first lieutenant in the 101st Airborne Division and had parachuted into Bastogne and in about an hour would be taking snapshots on his way to death. William Hellman, squad leader and second-in-command, was handsome and important and only a month before had made a parachute landing on the Ellipse between the White House and the Washington Monument. At the end he wished he had been a better Catholic, and men wept when they saw him still alive. Then there were the three survivors. R. Wagner ("Wag") Dodge, the foreman, gifted with his hands, silent on principle, and fastidious, who invented a fire and lay down in its ashes, lived only a few years more. Walter Rumsey and Robert Sallee, the other two who survived, spent years trying to forget the fire. Part of our story will be to find them and to bring them back to Mann Gulch with us to discover how well they remembered and forgot. In this part of the story there are living ghosts as characters, and the story doesn't come out quite the way either the ghosts or the non-ghosts expected.

But even in the memories of those who knew them, the dead Smokejumpers have a collective character. When you ask any one of them what he knew of any one of the dead, you always get the same answer, which is undoubtedly true: "He was a great guy." And when parents can bring themselves to speak of a son, they always say, "He was a wonderful boy." Of the fifteen who jumped on the fire, thirteen were between seventeen and twenty-three, and were still so young they didn't like the taste of hard liquor but drank beer, gallons of beer. Being that young, they were in good part what their training made them, and maybe with their girls—maybe especially with their girls—they acted mostly like Smokejumpers. Important in becoming a Smokejumper is learning how to act like one.

First-year Smokejumpers believed women were largely what second- or third-year Smokejumpers told them women were, and, having few chances to see women of the world while they were on the job in the woods, they had a very small body of fact to correct what they were told. Probably the "woman of the world" they knew best was at the nearest bar, about a four-and-a-half-mile walk down the road from their training base at Nine Mile, and sometimes, especially after returning from a jump, they wouldn't get to the bar until nearly 1:30 at night, which was closing time, but they would make the owner stay open until dawn. No matter the time, there was always this same tall dame on a stool waiting to accommodate herself to their beer. She was tall and silent, but after a couple of hours her trunk would begin to sway, and, as she finally toppled like a tree from her stool, the Smokejumpers would all stand and yell, "Timber." Then they would walk four and a half miles back to the base and be ready to jump.

A few such glimpses of women when mixed with verbal

images of them drawn by experienced jumpers must have left first-year jumpers picturing "women of the world" as part tree, probably part sheep, and certainly part deer, because almost without doubt they had branches and antlers.

What most first-year jumpers really knew about women was only one girl, the girl back home. She had been a junior in high school when he had been a senior. He took time off in the summer to be with her at the creamery picnic back home, and he took at least another weekend off during the summer just to be with her. Then they went on a long pack-trip together. She always carried at least forty pounds in her packsack, and they would stay out overnight. Like him and other great walkers, she walked slightly stoop-shouldered. At night after he returned to his job of smokejumping he would float into her dreams from the sky looking for a fire, and as he floated by always he stopped for a look at her. If he had to go on before he found her, she would wake up deeply disturbed, but, believe it or not, she never thought of him as tough.

It is hard to realize that these young men would be dead within two hours after they landed from parachutes no longer made of silk but of nylon, so they would not be eaten by grasshoppers.

————

So it was a young outfit, of necessity young as individuals and barely started as an organization. As individuals, they would soon go the way of prizefighters—all washed up when their reflexes began to slow by fractions of a second and when they no longer could absorb a beating and come back to win. Few have ever made it to the age limit of forty. As an organization, the outfit also was young, only nine years old in 1949, and some of those nine years were war years when the development of the Smokejumpers was slowed down. The war, though, had positive after-effects on the history of the Smokejumpers. For instance, of the fifteen who jumped on the Mann Gulch fire, twelve had been in the armed services and the other three had been too young to enlist.

In 1949, then, this was an outfit of great power and to us, but not to them, of some apparent weaknesses. Although modern graphs demonstrate the effectiveness of the Smoke-jumpers in carrying out their major purpose of putting out fires so fast they don't have time to become big ones, nothing initially—at least not before the Mann Gulch fire—made visible any weakness in adhering too exclusively to this purpose even when it would have made sense to enlarge it.

One danger of making almost a sole specialty of dropping on fires as soon as possible is that nearly all such fires will be small fires, and a tragic corollary is that not much about fighting big fires can be learned by fighting small ones. Small fires, remember, most frequently are put out with a shovel and an ax, to which, for the sake of the record, a Pulaski should be added, a Pulaski being a double-bitted ax with one of the bits made into a little hoe. As for big fires in the early history of the Forest Service, a young ranger made himself famous by answering the big question on an exam, "What would you do to control a crown fire?" with the one-liner, "Get out of the way and pray like hell for rain." Another weakness that might show up from a specialization of dropping on small fires in otherwise inaccessible country is that there aren't enough of them in a usual season to make them into a profession. In 1949, when there were big critical fires that could be approached on the ground, the Forest Service continued its original practice, as it was to do with the Mann Gulch fire, of going to the big towns—Spokane, Butte, Missoula, or Helena—and picking up what they could find sitting at bars or lined up outside employment agencies, maybe with one good pair of walking shoes among the whole bunch which one barfly passed to the other in the alley when it came his turn to be interviewed. In these early times, when Smokejumpers were not actually on fires, most of which were small, they were either at their base picking dandelions out of the lawn while waiting for their names to come to the head of the jump list or out in the woods on what were called "projects," building trail, stringing telephone wire, or thinning dense timber and not learning much if anything about what to do when a fire gets big enough to jump from one side of a canyon to the other.

The Smokejumpers are now the crack firefighters of the Forest Service, the shock troops. Whenever fires are critical, which practically always means big, that's where they are, from Missoula, Montana, to Minnesota to New Mexico to Alaska, and they don't care how they get there—by plane, bus, horse, or on foot, just so it is the fastest way. They are professional firefighters, for a living taking on fires of all sizes and shapes.

A Class C fire (10 to 99 acres) has a special place in this story, although as forest fires go it is no great fire. But many of this crew had never been on a fire as big as a Class C fire. Wag Dodge, the foreman of the crew and a Smokejumper foreman since 1945, had led one crew of Smokejumpers to one Class C fire in 1948 and to one Class D fire (100 to 299 acres) the year before but to no fire larger than that.

When the crew landed on the Mann Gulch fire, it was a C. Then suddenly it blew, and probably no one there had ever been on a "blowup" before. A blowup to a forest fire is something like a hurricane to an ocean storm. When 450 men finally got the Mann Gulch fire under control, it had burned forty-five hundred acres, between seven and eight square miles.

The primary purpose of the first Smokejumpers, then, was still primary to the Smokejumpers of 1949—to land on a forest fire in difficult or otherwise inaccessible country before suddenly the universe tried to reduce its own frame of things to ashes and charred grouse. When the Mann Gulch fire was first spotted from the plane, the pilot, the crew foreman, and the spotter sized it up as a fairly ordinary fire—they reported it was just a "ground fire" that had "crowned" in one place where it had already burned out. None of the three saw any "spot fires" around its edges, and that meant the fire had been advancing slowly on the ground and was not playing leapfrog by throwing small fires ahead of the fire's main front.

The words in quotation marks above and undoubtedly some that are not are those of firefighters, and we had better be sure of the meaning of these key words in the Basic English of firefighters so that when the tragic race between the firefighters and the fire begins it won't have to be stopped for definitions. It is not enough to know the word for this or that kind of fire; to know one fire is to see how what was dropping live ashes from a dead tree at the end of one afternoon by next afternoon had become one kind of fire after another kind of fire until it had become a monster in flames from which there was no escape.

Of the two main kinds of forest fires distinguished by their causes, man and nature itself, the Mann Gulch fire was a lightning fire, as 75 percent of the forest fires in the West are. Lightning fires usually start where lightning gets its first chance to strike—high up near the top of a ridge but slightly down its side where the first clump of dead trees stands, and the start of the Mann Gulch fire fits this description. The fire in the dead snag may drop live ashes for several days before starting a fire on the ground, for the ground near a mountaintop is likely to be mostly rocks with at best only a light covering of dead leaves, needles, or grass. But the lightning storm that started the Mann Gulch fire passed over the gulch on August 4, and by the end of the next afternoon on the hottest day ever recorded in nearby Helena thirteen Smokejumpers were dead.

Once started on the ground the lightning fire became simply a "ground fire," a term that includes most fires, and so ground fires are of many sizes, shapes, and intensities, and practically all man-made fires such as campfires and fires set to burn slash or brush but allowed to get away at least start as ground fires. A ground fire may become dangerous, even murderous, but most often it is just a lot of hard work to get under control. Until an hour before the end, that is what the Smokejumpers expected the Mann Gulch fire to be—hard work all night but easing up by morning.

The job of controlling most ground fires starts with the job of scraping a "fire trench" or fire-line around it or its flanks so as to force it onto rocks or open meadows. A fire trench or fire-line is some two to three feet wide, is made with a Pulaski and shovel, and is nothing more than the surface of the ground scraped down to mineral soil. Nothing flammable, such as fallen trees or hanging branches, can be left across it.

The chief danger from a ground fire is that it will become a "crown fire," that is, get into the branches or "crowns" of trees especially where the trees are close together and the branches interlace. So a crew has to be careful that a ground fire doesn't burn into a jack-pine thicket where the branches are close to the ground and can be set afire by low flames. But there is still a very different way for an ordinary-looking fire to explode. A fire doesn't always need flames to advance.

A fire may seem under control, burning harmlessly under tall trees with branches too high to be touched by ground flames, but the fire is burning with such intensity that most of the oxygen has been burned out of the air near it, which is heated above the point of ignition. If the wind suddenly changes and fresh air is blown in loaded with oxygen, then the three elements necessary for a fire are suddenly present in the lower branches—flammable material, temperature above the point of ignition, and oxygen. An old-timer knows that, when a ground fire explodes into a crown fire with nothing he can see to cause it, he has not witnessed spontaneous combustion but the outer appearance of the invisible pressure of a "fire triangle" suddenly in proper proportions for an explosion.

The crown fire is the one that sounds like a train coming too fast around a curve and may get so high-keyed the crew cannot understand what their foreman is trying to do to save them. Sometimes, when the timber thins out, it sounds as if the train were clicking across a bridge, sometimes it hits an open clearing and becomes hushed as if going through a tunnel, but when the burning cones swirl through the air and fall on the other side of the clearing, starting spot fires there, the new fire sounds as if it were the train coming out of the tunnel, belching black unburned smoke. The unburned smoke boils up until it reaches oxygen, then bursts into gigantic flames on top of its cloud of smoke in the sky. The new firefighter, seeing black smoke rise from the ground and then at the top of the sky turn into flames, thinks that natural law has been reversed. The flames should come first and the smoke from them. The new firefighter doesn't know how his fire got way up there. He is frightened and should be.

A fire-line, unless a river or a wide right-of-way on a trail is being used as a line, is not much good when a crown fire is off and running. It usually takes a "backfire" to stop a big crown fire, and the conditions are seldom right for the foreman to start one. He has to build piles of fast-burning twigs, shavings, or dried bunch grass in front of the main fire and, before starting his backfire, must wait until the wind blows back toward the main fire, and often it never does. When you fool with a backfire, you are really fooling with fire—you are counting on the wind to continue to blow your backfire toward the main fire. If the wind changes again and blows toward you, your backfire may only have given the main fire a fatal jump on you.

It's perhaps even more unpredictable if there isn't much of a wind to begin with, because a big crown fire can make its own wind. The hot, lighter air rises, the cold, heavier air rushes down to replace it in what is called a "convection effect," and soon a great "fire whirl" is started and fills the air with burning cones and branches which drop in advance of the main fire like the Fourth of July and start spot fires. The separate spot fires soon burn together, and life is trapped between the main fire coming from behind and the new line of fire now burning back toward it.

Then something terrible can happen. The space between the converged spot fires as they burn close to the main fire can become hotter than the point of ignition. If the convection effect or a change in the wind blows fresh oxygen between the two fires, suddenly replenishing the burned-out air, there can be a "blowup," although a blowup can be caused in still other ways. Not many have seen a blowup, even fewer have seen one and lived, and fewer still have tried afterwards to recover and record out of their seared memories exactly what happened. Later on in Mann Gulch we shall try to recreate a blowup seen by almost no one who lived to record it, and it might help as preparation if we turn briefly to the great pioneer in the science of fire behavior, Harry T. Gisborne, who was one of the first to observe and describe a blowup accurately.

In 1929 Gisborne was on what was up to then Montana's largest man-caused fire, the ninety-thousand-acre Half Moon fire in Glacier National Park (640 acres being a section or a square mile). As he says, measured "runs" show that even a big crown fire advances not much faster than a half-mile to a mile an hour. The blowup that Gisborne witnessed demolished over two square miles in possibly two minutes, although probably in a minute flat.

Returning two days later, he found the perfectly balanced body of a young grouse, neck and head "still alertly erect in fear and wonder," the beak, feathers, and feet seared away. Within a few yards was a squirrel, stretched out at full length. "The burned-off stubs of his little hands were reaching out as far ahead as possible, the back legs were extended to the full in one final, hopeless push, trying, like any human, to crawl just one painful inch further to escape this unnecessary death."

Although young men died like squirrels in Mann Gulch, the Mann Gulch fire should not end there, smoke drifting away and leaving terror without consolation of explanation, and controversy without lasting settlement. Probably most catastrophes end this way without an ending, the dead not even knowing how they died but "still alertly erect in fear and wonder," those who loved them forever questioning "this unnecessary death," and the rest of us tiring of this inconsolable catastrophe and turning to the next one. This is a catastrophe that we hope will not end where it began; it might go on and become a story. It will not have to be made up—that is all-important to us—but we do have to know in what odd places to look for missing parts of a story about a wildfire and of course have to know a story and a wildfire when we see one. So this story is a test of its own belief—that in this cockeyed world there are shapes and designs, if only we have some curiosity, training, and compassion and take care not to lie or be sentimental. It would be a start to a story if this catastrophe were found to have circled around out there somewhere until it could return to itself with explanations of its own mysteries and with the grief it left behind, not removed, because grief has its own place at or near the end of things, but altered somewhat by the addition of something like wonder—wonder, for example, because now we can say that the fire whirl which destroyed was caused by three winds on a river. If we could say something like this and be speaking both accurately and somewhat like Shelley when he spoke of clouds and winds, then what we would be talking about would start to change from catastrophe without a filled-in story to what could be called the story of a tragedy, but tragedy would be only a part of it, as it is of life.

2

The C-47 circled the fire several times before dropping the crew. The spotter, Earl Cooley, lay flat on the floor on the left side of the open door, with headphones on so he could talk with the pilot; and the foreman, Wag Dodge, lay on the right side of the door so that he and the spotter could watch the country together and talk without the crew hearing much of what they were saying. They were both experienced and good. In a later statement, Fred Stillings, administrator of the Smokejumping Project, said, "In my judgment they were the best available for the jobs assigned." And Cooley, remember, was one of the first two jumpers ever to parachute on a forest fire; later he was to become a successor to Stillings and administrator of the Smokejumping Project himself. Cooley is the only Smokejumper I ever heard say, "I don't know why, but I was never afraid to jump. It keeps others awake at night." He is a fine guy, but there is something not in him that is in us—he was always used on rescue work, and when some bush pilot didn't clear a ridge Cooley was sent to the top of the mountain to separate the pieces of the fuselage from the pieces of the pilot and put the latter pieces in a packsack and bring them back. After finishing a couple of his rescue stories, he says, "What the hell? If you work for the Forest Service, what do you expect of life?" As far as I know, he has never answered that question, except to himself.

Even lying on the floor of the plane, Wag Dodge was unwrinkled and handsome in his intentness. Like a fair number of men who can make anything with their hands, he was fastidious in his dress. His wife said, when he returned after a fire and his crew were black to inside their skins, he was a fashion plate coming down the steps of the plane. Even after his return from the Mann Gulch fire, where the fire had burned over him and he had stayed until all the bodies were recovered, he seemed as fastidious as ever until she got close enough to him to see the stain of tobacco juice at each corner of his mouth. To her knowledge, this was the only time he had ever chewed tobacco.

She and I have known each other off and on for most of our lives, and we have known the Blackfoot River, where her ranch house stands, even longer. "He said to me when we were married, 'You do your job and I'll do mine, and we'll get along just fine.'" Then she said to me, "I can't help you much. I don't know much about smokejumping, and I didn't know any of the Smokejumpers. We never talked about them, and he never invited them home." She added, "I loved him very much, but I didn't know him very well. If he said my red drapes were black, I would say, trying to keep myself intact, 'Yes, Wag, my red drapes are black.'"

His Smokejumpers didn't know much about him either, and he knew almost nothing about them. They knew he was one of the most experienced jumpers in the outfit, having been with the jumpers from one year after their founding. All told, he had been nine seasons with the Forest Service, and since 1945 he had been Smokejumper foreman. Actually, being so gifted with his hands may have been an indirect cause of the tragedy ahead of him.

The Smokejumpers have never had a fixed organization like the military, with the same squads and officers (in the case of the Smokejumpers, a foreman and a squad leader). The crew about to be dropped on Mann Gulch had never before worked on a fire under Dodge.

Since the cost of keeping separate crews intact during a hot fire season would be prohibitive, a list is posted of all the jumpers and "overhead" (foremen, squad leaders, and spotters), and when a man has been on a fire he is dropped to the bottom of the list and has to work his way back up. No one knows who or how many will be called next, especially since the number of jumpers dropped on a fire can vary from two men to several planeloads. You don't have to be an administrative genius to see in this organizational scheme of things the possibility of calamity in a crisis.

The Forest Service, realizing some of the danger that might arise in a clutch when men whose lives depended upon each other did not know each other, had instituted a three-week training course at the beginning of each fire season in which crew members and overhead worked together. Whether this course produced much familiarity between crew and overhead is a question not worth asking about Mann Gulch: Wag Dodge was so good with his hands that in the spring of 1949 he had been made "barn boss" in charge of the maintenance of the base and so had been unable to take part in the course.

Another question is sure to be asked before the fire in Mann Gulch is forever cold. Was the three-week training course adequate to the emergencies Smokejumpers must meet? The Smokejumpers were still a young outfit that hadn't figured itself out yet, and it still had to fight for a budget. It didn't have money, for instance, to keep many of the jumpers around the base unless there were a lot of fires burning, so when things were quiet they were sent out on projects, which might mean that a jumper would find himself on a crew building trail, and that's about as unenlightened work as you can get in the woods or anywhere else.

One thing, though, that the Smokejumpers have been

taught well from the beginning is pride, and you can't be much of a firefighter without it, and without it you certainly can't jump out of a plane when you get sick every time you leave the ground, as some jumpers do. From the beginning one of the great realities of the Smokejumpers is its romanticism. There's nothing wrong with romanticism, except that sometimes it isn't enough. They were jumping into one of the roughest pieces of country Lewis and Clark had seen on their long, long journey to the Pacific, one that does not pardon weakness on a hot afternoon of a burning summer.

The foreman who lay on the right side of the door of the C-47 was in many ways all that a foreman in the Smokejumpers should be. A foreman is supposed to do everything his men do and do it first and better. Dodge could do all that with a margin to spare—he could do things with his hands approaching artistry. He was impeccable and inflexible. But he didn't tell his wife much, and he also didn't know most of his crew, even by name, and they didn't know him, except by name.

They spotted the fire and circled it once. In their statements to the Forest Service Board of Review after the fire, the pilot, the spotter, and the foreman agree that, when first spotted, the fire covered between fifty and sixty acres and was burning on the ridge between Mann Gulch and Meriwether Canyon and partway down the Mann Gulch side, threatening but not as yet burning into Meriwether. Even from the air, the fire's brief history could be read—it had started on the Mann Gulch side (as it turned out, by lightning the afternoon before), had made a run for the top, and had burned hot enough to crown on the way up. Then the crown had burned out, never burning hot enough to start any spot fires.

On the ridge where the fire seemed to be most active it was burning downhill into a saddle where the fuel, chiefly

grass, was thin. Besides, a fire generally burns slower downhill than up, ashes rolling downhill more slowly than flames
rise up it. It was 2:30 when the plane left the Smokejumper
base in Missoula and around 3:10 in the afternoon when the
fire was first spotted from the plane, late enough for the wind
and the temperature to start easing off, and, although August
5, 1949, was in the middle of a heat wave and the official
temperature in Helena, twenty-five miles away, was ninety-
seven degrees Fahrenheit, the hottest day on record, and the
fire danger rating was high at 74 with 100 as maximum, yet
all three observers as they circled the fire regarded it as routine. A spotter always has the right, although he doesn't like
to exercise it, of returning to the base without jumping his
men if he thinks the wind too strong (twenty miles or more
an hour) or the terrain or the fire too dangerous. Cooley and
Dodge took a good look at the fire, thought it more or less
routine, and figured the crew would have it under control
by ten o'clock the next morning. They referred to the smoke
leaking out of the ridge as merely a cauliflower, although
soon it was more like a leak in a lobe of the brain of the
universe. On the second swing around they were trying to
pick out a landing area. They were also trying to get some
idea of the general lay of the land, which was unfamiliar
to them. They had been sent from their base in Missoula
without maps, on the poor excuse that there would probably
already be ground crews on the fire with maps. Even if there
had been, it would have made no difference.

———

The fire was located in the "Gates of the Mountains" wild area
(roadless area) just east of the Missouri River, some twenty miles
north of Helena . . . at a point near the top of the ridge between
Mann and Meriwether gulches. The general area is steep and jag-

ged on the Meriwether side and is said to be one of the roughest
areas east of the Continental Divide.

From the official *Report of Board of Review*, Mann Gulch fire, Helena National
Forest, August 5, 1949.

Mann Gulch is a dry gulch two and a half miles long that runs
into the lower end of the spectacular stretch of the Missouri
River called the Gates of the Mountains by the first white
man who entered them, Captain Meriwether Lewis, when on
July 19, 1805, he camped his party at the mouth of the gulch
now bearing his first name. Immediately downriver from
Meriwether Canyon is Mann Gulch, where the fire started
near the top of the ridge between the two gulches, and almost
immediately downriver from these two gulches the Gates
open to the plains.

"From the singular appearance of this place I called it
the gates of the rocky mountains," Captain Lewis said in his
journals. Its singular appearance makes it a fitting backdrop
for early and everlasting drama in which nature plays the
leading role. If you are coming upstream from yellow flat
plains, as Captain Lewis and Captain Clark had been for over
a year, you can observe even at a great distance how there
is something about mountains that hates to be plains. Far,
far ahead are the mountains black with the haze that makes
mountains look from the plains as if they were clouds of
smoke from a great forest fire. As they and you come closer,
the haze of the mountains breaks apart and reluctantly allows the yellow plains a final appearance. This is literally the
way it was in Mann Gulch before the fire burned it out in a
matter of minutes. It was the place in the Gates where the
struggle between mountains and plains came face to face—
below Mann Gulch belongs to the plains, upriver to the
mountains and timber. Mann Gulch itself where the grave

markers are was yellow with tall grass. The differences are not only scenic—there are differences between the behavior of grass and timber fires, and the differences can be tragic if firefighters don't know them.

The Smokejumpers were on their way to a blowup, a catastrophic collision of fire, clouds, and winds. With almost dramatic fitness, the collision was to occur where vast geological confrontations had occurred millions and millions of years ago—where old ocean beds, the bottoms of inland seas, were hoisted vertically by causes too long ago to be now identified and were then thrust forward by gravity into and over other ocean beds, cracking and crumbling them and creasing them into folds and creating a geological area called in the subdued language of scientists the "Disturbed Belt," a belt that includes in its geological history much of not only northwestern Montana but western Alberta and eastern British Columbia.

The "Disturbed Belt" in turn is a loosely tied part of a much larger geological formation scientists call the "Overthrust Belt." This overthrust formation not only includes the front or face of the Rockies from western Alberta and eastern British Columbia on down through Glacier Park and northwestern Montana (the "Disturbed Belt"); it extends in both directions to northern Canada and Alaska and all the way to central Mexico. Prior to the formation of this gigantic extrusion some 150 million years ago, large portions of the western margin of our continent, which then lay several hundred miles east of our present Pacific Coast, were covered by deep layers of sedimentary rocks, limestone and sandstone, deposited there by transient inland oceans which must have been something like Hudson Bay. As the western continent was raised, squeezed, and compressed, great slabs of sedimentary layers slid over each other inland or eastward for a distance varying from a few miles to a hundred miles or more.

The present cliffs in the Gates of the Mountains are the rearings and collisions and roarings of the bottoms of oceans as they stood up like sea beasts struggling to prevent anything from finding a way around them. The cliffs on the sides of each canyon are bases to arches that once rested on the cliffs, as is proved by the matching strata on the opposite cliffs, but the key to the arch that once joined the cliffs has gone off somewhere and been replaced by the eternal arch of Montana sky.

In the Gates of the Mountains there have been many blowups. Now there are many rattlesnakes and nothing more fragile than mountain goats, themselves tougher than the mountains they disdain, although at a distance they are white wings of butterflies floating up and down and sideways across the faces of fragments of arches and cliffs, touching but never becoming attached to them.

When the Missouri escapes at the Gates from around a bend or from under a mountain it is still clear, but almost immediately after entering the plains it turns yellow like the plains and from then on there are plains and plains and plains, yellow plains parted only by a yellow river.

Do not be deceived, though, by the scenic beauty of the Gates of the Mountains into believing that the confrontations and terrors of nature are obsolescences frozen in stone, like the battles of satyrs in Greek bas-relief, remnants of mythology and witnessed if ever by dinosaurs and now only by seismographs. It is easy for us to assume that as the result of modern science "we have conquered nature," that nature is now confined to beaches for children and to national parks where the few remaining grizzly bears have been shot with tranquilizers and removed to above the timberline, supposedly for their safety and our own. But we should be prepared for the possibility, even if we are going to accompany modern firefighters into Mann Gulch, that the terror of the universe has not yet fossilized and the universe has not run out of blowups.

Yet we should also go on wondering if there is not some shape, form, design as of artistry in this universe we are entering that is composed of catastrophes and missing parts. Whether we are coming up or down the Gates of the Mountains, catastrophes everywhere enfold us as they do the river, and catastrophes may seem to be only the visible remains of defunct happenings of millions of years ago and the Rocky Mountains only the disintegrated explosions that darkened skies also millions of years ago and left behind the world dusted with gritty silicon. At least I should recognize this as much the same stuff as the little pieces of glass which in 1980 Mount St. Helens in Washington sprinkled over my cabin in Montana six hundred miles away, and anyone coming down the Gates of the Mountains can see that the laminations of ocean beds compressed in the cliffs on one side of the river match the laminations in the opposite cliffs, and, looking up, can see that an arch, now disappeared into sky, originally joined both cliffs. There are also missing parts to the story of the lonely crosses ahead of us, almost invisible in deep grass near the top of a mountain. What if, by searching the earth and even the sky for these missing parts, we should find enough of them to see catastrophe change into the shape of remembered tragedy? Unless we are willing to escape into sentimentality or fantasy, often the best we can do with catastrophes, even our own, is to find out exactly what happened and restore some of the missing parts—hopefully, even the arch to the sky.

Even on the first run over the fire, all pertinent pieces of the plane and its universe began to fall into place and become one, preparatory to the jump—the crew, the overhead, the pilot, the airplane, the gulch, the fire in it, and the sky between, all readying themselves for the act. Jumping is one of the few jobs in the world that leads to just one moment when you must be just highly selected pieces of yourself that fit exactly the pieces of your training, your pieces of equipment having been made with those pieces of yourself and your training in mind. Each of the crew is sitting between the other's legs, and all this is leading to a single act performed between heaven and earth by you alone, all your pieces having to be for this one moment just one piece. If you are alive at the end of the act, it has taken about a minute—less, if you are not alive. The jump is that kind of beauty when everything has to be in perfect unison in order for men to commit themselves to what once done cannot be recalled and at best can be only slightly modified. It becomes the perfectly coordinated effort when a *woof* is heard on earth as the parachute explodes open within five seconds after the jumper steps into the sky. If it's more than five seconds, a handle has to be pulled to release the emergency chute.

The pilot now was circling to see how close he could get the plane and the jumpers to the fire. The circles became closer by becoming smaller and nearer to the ground. Sometimes a Smokejumper pilot gets so near to the ground after the crew jumps but before he can pull his plane out of the face of a mountain that he returns to the base in Missoula with evergreen boughs in his landing gear. That is also beautiful, but mathematically can't happen often.

It was the time now for the drawing together of the overhead and the pilot. The spotter and the foreman lay on the floor with only the open door between them; the spotter and the pilot were joined together by earphones. The pilot was Kenneth Huber, a good one. He had been flying for the famous Johnson Flying Service for four years and during the war had transported paratroopers.

The Johnson Flying Service did all the flying for the Smokejumpers in Missoula on contract and was as much a western legend as the jumpers themselves. In the Northwest, Bob Johnson, the owner, was a kind of Paul Bunyan of the air. Huber told Cooley over the earphones that his altimeter showed the plane had dropped a thousand feet in a few minutes and that because of the suction of air in the gulch he was going to jump the men above the ridge—at two thousand feet instead of the customary twelve hundred. Cooley knew as a result that the crew and cargo on landing would be scattered more than usual, and this information affected Cooley's and Dodge's choice of a landing area.

If Bob Johnson had been piloting his own plane, he probably would have taken it into the gulch and arrived back in Missoula physically exhausted from lifting his plane out of the cliffs, and the plane would have been ornamented with Christmas boughs.

On their first pass over the fire, the pilot, the spotter, and the foreman were already looking for a possible jump site, but the job of picking one was primarily the spotter's. The pilot's job, since he generally uses instruments when jumping his men, is to report what he can see on his instruments, and he had already reported considerable wind turbulence in the gulch. Both Cooley and Dodge, looking through the open door, immediately noticed a possible jump area, right on top of the ridge and right in front of the fire on its upgulch side. But almost immediately they said "no" and shook their heads in case "no" could not be heard across the open door. Naturally, they were trying to drop their men close to the fire—but without endangering them or their equipment, and a change in the wind, which was blowing upgulch from the river, so close to a fire front might have been the end of both. They were starting with the knowledge from the instruments that the wind had dropped the plane a thousand feet in minutes, and to this they could add their own stored-up knowledge that the top of a mountain is a world particularly devoid of equilibrium. As Cooley, the most experienced of the overhead, said later in explaining his rejection of this landing site, if you have wind turbulence to start with, you should know ahead of time that there will be even more of it at the top of a mountain and that, at the top, one side will have an upwind and the other side a downwind. Those who died later died near the top of a mountain in the upwind.

On the next pass Cooley selected a jump area near the head of the gulch on its north side where "the slope gradually goes off into the bottom and your jumper more or less hits equilibrium." Cooley later told the Forest Service's Board of Review that he estimated the jump area to be "a strong half mile from the closest point of the fire" and five hundred feet below it in elevation—not only below it but on its flank and, important also, with few trees and rocks. Dodge finally accepted the site, although first objecting because a helicopter couldn't land there in case injured men had to be brought out.

So they all tried to think of everything, but the pilot thought primarily of his instruments, and the foreman thought primarily of his crew, and the spotter thought primarily of everything and made the decision.

Then the crew began to stir. They were sitting straddle-legged on the floor, their backs to the cockpit, each man fitting snugly between the legs of the man behind him so that all sixteen jumpers and their equipment could be packed into the plane. They were almost literally one body—the equipment of each man next to him where the seat had been and each man between the legs of the man behind. Since the beginning of the flight, the assistant spotter, Jack Nash, had been checking their equipment. Now the men stirred to check themselves, figuring they had better do their own checking since they were going to have to do their own jumping.

On the next pass over the fire the assistant spotter stood by the side of the open door and dropped the hunter orange drift chutes, and on the next circle the spotter estimated the distance and direction the wind had blown the chutes so he could tell how far ahead of their target he should drop the jumpers.

The first "stick" stood up, a stick being the number of men, usually three or four, who are jumped on each run over the landing area. They stand in front of the open door, one behind the other, the front man with his left foot forward. They are closer together than ever. The man behind the first man stands with his right foot forward so that after the first man jumps the second man can make one step forward with his left foot and be where the first man was.

It is the assistant spotter's job to snap the jumper's static line to a rod in the roof of the plane. The other end of the static line is snapped to his parachute. It is twelve feet long and, if all else goes well, will automatically open the jumper's parachute after he has fallen twelve feet. So as the moment for the jump approaches, the men and the plane get closer and closer together. Some jumpers won't allow the assistant spotter to snap their static line to the plane—they do it themselves. They also have to be careful when they jump not to hang themselves in their own lines. One has. On the next pass they started jumping. The foreman jumped first.

Nearly every jumper fears this moment. If he continues to miss sleep because of it, he doesn't tell anybody but he quits the Smokejumpers and joins up with something like the crew that makes trails. Whatever he tries, it is something close to the ground, and he never tries jumping again because it makes him vomit.

Fear could be part of the reason they were jumping only fifteen men on this day—one had become sick on the flight over. Although he was an experienced jumper, his repressions had caught up with him and he had become ill on each of his flights this season and had not been able to jump. This was a rough trip, and after he had vomited and crawled out of it and his jump suit, he must have made his decision. When he landed back in Missoula, he resigned from the Smokejumpers.

It was a record temperature outside and the air was turbulent, so much so that Sallee once told me that they were all half sick and trying to be in the first stick to jump and get on the ground. But, weather aside, it was hard to know on what day this or that good man had built up more anxiety than he could handle, and at the last minute on this day this crew of fifteen was jumping four sticks of 4-4-4-3. On the ground, however, the crew was to pick up another firefighter who had been fighting the fire alone, so when the showdown came the crew was again sixteen.

The fear of the jumpers is a complicated matter, because in some ways a part of each of them is not afraid. Most of them, for instance, believe that God is out there, or a spirit or a something in the sky that holds them up. "You wouldn't dare jump," they say, "if it was empty out there." Also they say, "Why be afraid? You are jumping in a parachute, and the government made the parachute, didn't it?" This is connected with their thinking that guys who hang glide from the tops of the big mountains surrounding Missoula are crazy. "They're crazy," the Smokejumpers say. "They don't have a government parachute." So in some strange way they think they are jumping on the wings of God and the government. This does not keep them from worrying some nights—maybe every night—before they jump, and it does not keep some of them from vomiting as they are about to jump.

Understandably, Smokejumpers have an obsession about their equipment. Although they change from one fixation to another, equipment is nearly always somewhere on their mind, and, as they get close to the jump, equipment is about all that is on their mind. They know they are about to live or die on a man-made substitute for wings furnished by the government. They start saying to themselves, as if it had never occurred to them before, "What the hell does the government know about making a parachute that will open five seconds after it starts to fall? Not a damn thing. They just farm it out to some fly-by-night outfit that makes the lowest bid." As the jump nears, their general fears focus on what seems the least substantial and the most critical piece of their equipment—the static line that is supposed to jerk the parachute open with a *woof* twelve feet after it drops from the plane.

The attention the jumper has to pay to the elaborate and studied ritual of jumping helps to keep his fears manageable. He stands by the spotter lying on the left of the door, who holds the jumper by the left foot. The next signs are by touch and not by word—the whole flight is made with the door open, unless it is going to be a very long one, so words can't be trusted in the roar of the wind. Using the sill of the open door as a gunsight, the spotter waits for the landing area to appear in it and next allows for the wind drift. The spotter then says "Go," or something like that, but the jumper doesn't step into the sky until he feels the tap on the calf of his left leg, and in his dreams he remembers the tap. With the tap he steps into the sky left foot first so that the wind drift will not throw him face-first into the plane's tail just to his left. He leaves for earth in the "tuck position," a position somewhat like the one he was in before he was born. This whole business of appearing on earth from the sky has several likenesses to nativity.

The jumpers are forced into this crouched, prenatal moment almost by the frame of things. The jumper, unlike the hang glider, is not up there for scenic purposes. He comes closer to plummeting than to gliding. He is to land as close as possible to the target the spotter has picked, and all the jumpers are supposed to do the same so no time will be lost in collecting and piling their stuff in the same pile and being off to the fire. In order to drop as straight as possible, the jumpers originally would stand straight up in front of the door of the plane and the spotter would say, "Do you see the jump spot down there?" But if the jumper was a new man, the spotter wouldn't look to see if the jumper was seeing. He knew the jumper would be standing rigid with his eyes squeezed shut, looking as if he were looking at the distant horizon. But the spotter, needing to be sure that at least he was heard, would ask again, "Do you see the jump spot down there?" And the new jumper, frozen on the horizon, would say, "Yes, sir." Then he would get the tap on the left leg, but before he could jump he had to crouch in the tuck position because the favorite plane of the early Smokejumpers was the Ford TriMotor that had just a small opening for a door. So it was more or less the frame of things that forced a Smokejumper to be born again as he jumped.

His whole flight to the ground takes an average of only a minute. This minute is about the only moment a Smokejumper is ever alone, and it is one of the most lonely moments in his life. A Smokejumper never is sent alone to a fire; the minimum number is two; at their base Smokejumpers live in their dormitory with roommates or, if they live in Missoula, with their families; at night they are with their girls and often with other Smokejumpers who are with their girls, and if they get into a fight at a bar they are immediately supported by these other Smokejumpers. For the eternity of this one minute Smokejumpers are alone. It is not that they lose faith in God for that moment. It is just that He is not there anymore or anywhere else. Nothing is there except the jumper and his equipment made by the lowest bidder, and he himself has thinned out to the vanishing point of being only decisions once made that he can't do anything about ever after.

The moment the jumper starts falling is umbilical; he starts by counting, putting "one thousand" in front of each number to slow each count to a second. If he gets to "one thousand five," he knows he is in trouble and pulls the handle that releases the emergency chute on his chest. If, however, his umbilical relation to the plane is properly severed by his twelve-foot static line, his regular parachute explodes, the *woof* vibrates in the rocks below, and his feet are thrown over his head. So it is to be born in the sky—with a loud noise and your feet where your head ought to be. So it is to be born in the sky with a loud noise—the moment you cease to be umbilical you become seed, blown by the wind. It is very lonely for a young man to be seed in the wind. Although you are seed, the sky still seems like the womb and you as seed are blown around the sky's interior parts until you light on the top of a tree or hard rocks or grass, the grass often being only a cover for hard rocks. If you land on the top of a tree, you are probably lucky, especially if you have a long rope in your pocket by which you can let yourself down to the rocks—but only a small percentage when they touch earth land on the tops of trees. Try as they may to avoid landing on rocks, many do. Landing smoothly from the sky does not come naturally to man.

As in life generally, it is most common to land in grass that thinly covers very hard rocks. If a jumper lands on flat ground at all, it is something like jumping off the roof of an automobile going twenty-five miles an hour, and in 1949 he finished his jump by taking the "Allen roll," landing sideways, with the right side from the hip down taking the shock, the upper part of the body continuing to pivot to the right until the body falls on its back and then rolls over on its knees. As a jumping instructor once said, the roll is to spread the pain all over the body.

So it is to appear on the earth from the sky. It is not surprising, considering the punishment the jumper takes at both ends of the jump, that no big man can be a Smokejumper, and we have to remind ourselves from time to time that, although we keep saying "men," most of them are still close to boys and that they are not very big boys. Most of the seventeen or eighteen thousand visitors a year at the Smokejumper base in Missoula, having heard, possibly from the Smokejumpers themselves, that the Smokejumpers are the Forest Service's best, expect to see the Minnesota Vikings professional football team practicing outside their dormitory, but instead they see teams of fairly ordinary-looking boys playing volleyball, their sizes ranging from five feet four to six feet two, with a maximum weight in 1949 of 190 pounds. The name of the game is not important to Smokejumpers. The competition is. In the Smokejumpers they don't recruit losers or big men, who don't seem to be made to drop out of the sky.

This was a fairly rough landing. Sallee lit in a lodgepole, his feet just off the ground, but none of the rest of them were lucky enough to break their fall. They rolled through rocks, although only Dodge was injured. Hellman and Rumsey came to help him and found him with an elbow cut to white bone, the cut somehow self-sealed so that it did not bleed. They bandaged the elbow, and Dodge said only that it was stiff, and the next day he said only that it was stiffer.

They crawled out of their jump suits that made them look part spacemen and part football players. In 1949 they even wore regular leather football helmets; then there was wire mesh over their faces, the padded canvas suit (with damn little padding), and logger boots. They tagged their jump suits and stacked them in one pile. Their work clothes, unlike their jump suits, were their own, and they were mostly just ordinary work clothes—Levis and blue shirts, but hard hats. None in this crew appeared in white shirts and oxfords, although Smokejumpers have appeared on fires in their drinking clothes when there has been an emergency call and they have been picked up in a bar, and a jumper is quite a sight in a white shirt and oxfords after he has been on a fire for three or four days and had a hangover to start with.

Then the plane began to circle, dropping cargo. It was being dropped high and was scattering all over the head of the gulch. Because the cargo had been dropped at two thousand feet instead of the customary twelve hundred so the pilot would not have to take his plane close to the ridgetops in the heavy winds, the men had to collect the cargo over at least a three-hundred-square-yard area. In those days the bedrolls were dropped without benefit of parachutes and popped all over the landscape, some of them bouncing half as high as the trees. The parachutes were made of nylon because grasshoppers like the taste of silk. In a modern tragedy you have to watch out for little details rather than big flaws. By the end, every minute would count, but it took the crew some extra minutes to collect the cargo because it was so scattered. Suddenly there was a terrific crash about a quarter of a mile down the canyon from the landing area. It turned out to be the radio, whose parachute hadn't opened because its static line had broken where it was attached to the plane. Another detail. The pulverized radio, which had fallen straight, told the crew about how far downgulch from the landing area they had been jumped, so the spotter must have been allowing for about a quarter of a mile of wind drift. It also told them something else—that the outside world had disappeared. The only world had become Mann Gulch and a fire, and the two were soon to become one and the same and never to be separated, at least in story.

They finished collecting and piling up their cargo. Dodge estimated that the crew and cargo were dropped by 4:10 P.M. but that it was nearly 5:00 before all the cargo had been retrieved.

Dodge made the double L signal on the landing area with orange sleeves, signaling to the plane, all present and accounted for. The plane circled twice to be sure and then headed for the outside world. It headed straight down Mann Gulch and across the glare of the Missouri. It seemed to be leaving frighteningly fast, and it was. It had started out a freight train, loaded with cargo. Now it was light and fast and was gone. Its departure left the world much smaller.

There was nothing in the universe now but the terminal glare of the Missouri, an amphitheater of stone erected by geology, and a sixty-acre fire with a future. Whatever the future, it was all to take place here, and soon. Of the Smokejumpers' three elements, sky had already changed to earth. In about an hour the earth and even the sky would all be fire.

They could see the fire from their cargo area, at least they could see its flank on the Mann Gulch slope, and even at five o'clock they were not greatly impressed. Rumsey didn't think any of them regarded it as dangerous, although he did think it would be hard to mop up because it was burning on steep and rocky ground.

Then they heard a shout from the fire, but it was impossible to distinguish the words. The crew had been led to believe before they left Missoula that there would be a ground crew on the fire (hence, their having no maps), so Dodge told the squad leader, Bill Hellman, to take charge of the men and see that they ate something and filled their canteens while he himself took off for the fire to find out who was on it.

They spent only about ten minutes at the cargo area before they started tooling up. Sallee and Navon carried the saws; the rest were double-tooled. They thought they were going to work. Actually they were leaving an early station of the cross, where minutes anywhere along the way would have saved them.

3

Since their tools had better fit our hands if we are going to a fire, we should try them on here and see how they would have been used if the fire had been reached while it was at its present size of about sixty acres. At that size it is doubtful that the crew would have tried to hit it on its nose—it is dangerous business to attack a good-sized fire straight on.

Instead, they would have started flanking it close to its front and tried to steer it into some open ground, some stretch of shale or light grass where the fire would burn itself out or burn so feebly that it would be safe to take on directly. It's a ground fire of this size that, as suggested earlier, is brought under control by digging a fire-line around it, a shallow trench two to three feet wide scraped deep enough to expose mineral soil. All dead leaves, needles, even roots are removed so that nothing can burn across it. If any dead trees lie across it, they also must be removed and likewise any standing trees with low branches that the fire might use to jump the line. To put a fire "under control" is to establish and then hold such a line around it, especially around the part of it that is most likely to advance. What follows is called "mopping up," working back from the fire-line into the interior of the fire, digging shallow graves and dropping still-burning trees into them, and of course burying everything on the ground that smokes.

The tools that perform these two operations are, with one exception, those that have done most of the hard work of the world—axes, saws, and shovels.

Sallee says he was single-tooled and was carrying a saw, and Navon started with the other saw, which he soon traded off to Rumsey, who was carrying the heavy water can. Power saws, of course, were already invented, but those early ones were mechanical monsters; it took a whole crew just to crank one, so they were of no use to the Smokejumpers until well into the 1950s. The crew's two saws would have been two-man handsaws, and in making a fire-line would have been used to cut trees lying across the line or standing too close to it. In mopping up, they would have been used to drop the burning snags.

The not-always-clear references to tools by the two surviving crew members indicate that besides these two handsaws the crew had two or three shovels and eleven or twelve Pulaskis. Laird Robinson, who when I first met him was information officer at the Smokejumper base in Missoula, says that number sounds about right for a crew of sixteen at that time.

Even the numbers show that the Smokejumpers' tool of tools was the Pulaski. It was the forest firefighters' one invention, primitive but effective, invented strictly for firefighting. It was even named after the Forest Service's most famous firefighting ranger, Edward Pulaski, who in 1910, when many thought the world was ending in flames, put a gunnysack around his head and led forty-two half-paralyzed men through smoke to a deserted mining tunnel that he remembered. The cold air rushed out of the tunnel and was replaced by heat so intense it set fire to the mining timbers. Pulaski kept the fire in the tunnel under control by dipping water with his hat from a little stream that went by the mouth of the shaft, and he had enough control over his men to make them lie flat with their mouths on the ground. He was badly burned and finally passed out, and from time to time they all fell unconscious. But all recovered except five men and two horses.

The Pulaski is a kind of hybrid creation, half ax and half hoe. I remember the first one I ever used, an early, handmade one, nothing more than a double-bitted ax with one bit left on and a little hoe welded to where the other ax-bit had been. Even after all these years the Pulaski is still the tool for digging fire-lines. A little hoe goes deep enough because its job is to scrape the stuff that would burn off the surface of the ground. So the hoe makes the line; the ax-bit chops little trees or shrubs along the line that might let the fire jump across, and it has other uses, such as chopping roots. When the foreman ends his first lesson to his trainees on how to use a Pulaski, he says, "For the next couple of hours, all I want to see are your asses and your elbows."

Behind the crew with the fast Pulaskis come a couple of men with shovels, who clean out and widen the fire-line, and, of course, in the mopping-up operations, shovels are all-important in making shallow graves and burying whatever is still smoking.

The crew strung out on the trail. Those with the unsheathed saws were behind because the long teeth and rakers of the saws make them hard to carry and dangerous to follow too closely; most of the double-tooled men were carrying Pulaskis and for the second tool either a shovel or a water canteen or a first-aid kit or a rattlesnake kit. The flank of the fire was in plain view only half a mile across the gulch. Although from the cargo area its most advanced front on top of the ridge was not visible, they had seen it from the sky and remembered that on top of the ridge it was burning slowly downhill into a saddle. They had no trouble guessing what they would be doing ten or fifteen minutes from now when they caught up to their foreman and the fire. He would line them out on both the Mann Gulch and Meriwether flanks to make fire-lines that would keep the fire from spreading farther down either canyon and so limit its advance to the top of the ridge where, forced into the saddle and light grass, it would be easy to handle. Dodge would space the men with Pulaskis about ten to fifteen feet apart, depending upon the ground cover, and they wouldn't raise their heads until they had caught up to the man in front of them. Then they would tap him on the leg with a Pulaski and say "Bump." If the two men right behind had also finished their stretch, they would say "Bump Three." To a Smokejumper, "Bump" is a musical word if he is the one who sings it out.

When Smokejumpers work next to a regular crew of Forest Service firefighters, they take pleasure in leaving them bruised with "bumps."

As the crew started for the south side of the gulch, they had it figured out before they even had an order. They would work all night establishing a line around the fire. From then on, it would depend. The Smokejumpers couldn't be touched when it came to getting a line around a fire, but they usually didn't win medals in mopping it up. They were all in the business for money—the forestry school students, the fancy M.A., M.D., and Ph.D. students, and especially the jump-happy boys who hoped to make enough money in the summer to shack up all winter in Honolulu. So there was no use putting a little fire out of its misery too soon when you would be paid overtime.

The crew started up the side of the gulch toward the fire. It was about five o'clock. The next day a wristwatch of one of the boys was found near his body. Its hands were permanently melted at about four minutes to six. This must come close to marking the time when it was also over for most of the others. So there were about fifty-six minutes ahead of them, time to do only a little thinking, and undoubtedly only a little is all they did.

It is not hard to imagine what was in their heads. They knew they were the best and they were probably thinking at least indirectly about being the best, sizing up the fire ahead as a kind of pushover. They thought of what they were in as a game and they were the champs and the fire didn't look like much competition. They already had developed one of the best ways of facing danger in the woods, the habit of imagining you are being watched. You picture the mountainsides as sides of an amphitheater crowded with admirers, among whom always is your father, who fought fires in his time, and your girl, but even more clearly you can see yourself as champion crawling through the ropes. You would give this smalltime amateur fire the one-two, and go home and drink beer. It was more than one hundred degrees on that open hillside, and all of them were certainly thinking of beer. If anything troubled them, it was the thought of some guy they had tangled with in a Missoula bar who they were hoping would show up again tomorrow night. And each boy from a small town such as Darby, Montana, or Sandpoint, Idaho, was undoubtedly thinking of his small-town girl, who was just finishing high school a year behind him. She had big legs and rather small breasts that did not get in the way. She was strong like him, and a great walker like him, and she could pack forty pounds all day. He thought of her as walking with him now and shyly showing her love by offering to pack one of his double-tools. He was thinking he was returning her love by shyly refusing to let her.

The answer, then, to what was in their heads when they started for the fire has to be "Not much."

Like the Frontier Cavalry, the Smokejumpers didn't kill themselves off at the start of a march. They loosened up for about a quarter of a mile downgulch and then began to climb toward the fire, but they hadn't climbed more than a hundred yards before they heard Dodge call to them from above to stay where they were. Shortly he showed up with Jim Harrison, the recreation and fire prevention guard stationed at the campground at the mouth of Meriwether Canyon. Harrison had spotted the fire late in the morning while on patrol duty, returned to Meriwether Station, and tried unsuccessfully to radio both Missoula and Canyon Ferry Ranger Station outside Helena at 12:15, ten minutes before the fire was first officially reported by the lookout on Colorado Mountain, thirty miles away. After he had tacked a sign on the station door, "Gone to the fire. Jim," he again had climbed the fifteen-hundred-foot precipice between Meriwether Canyon and Mann Gulch, and had been on the front of the fire alone until Dodge found him around five o'clock. He had tried to do what he was supposed to do—stop the fire from burning down into scenic Meriwether Canyon. Meriwether Canyon is a chimney of fifteen-hundred-foot precipices and pinnacles. In minutes it could draw flames through the length of its funnel and be heat-cracked rocks forever after. It is one of America's tourist treasures, and Harrison had fought to save it. Later, two sections of fire-line that he had scraped with his Pulaski were found burned over at the top of the ridge between Meriwether and Mann. His tracks were there too, burned over.

Harrison was known to many of this crew because he had been a Smokejumper himself the summer before in Missoula, and ironically had switched to patrol duty and cleaning up picnic grounds to please his mother, who was afraid smokejumping was dangerous. Now here he was with Dodge and this crew of Smokejumpers on its mission of August 5,

1949, and he might as well have run into General Custer and the Seventh Cavalry on June 25, 1876, on their way to the Little Big Horn.

In addition, as recreation and patrol guard he could not have been in as good physical shape as the jumpers—the forest supervisor's description of his job makes clear that primarily he was a "recreation guard," keeping the public grounds and facilities at Meriwether Landing tidy for the tourists, and only upon special assignment was he to get into patrol and fire prevention work. As the supervisor told the Board of Review, Harrison had made only one patrol before August 5. Realizing he was in need of exercise, Harrison would hike up to his patrol point on his days off, but he couldn't have been in shape to keep up with the jumpers if the going got tough. And yet, that day he had twice climbed the perpendicular trail to the top of the ridge between Meriwether and Mann and fought fire alone for four hours while the Smokejumpers had done nothing but jump and walk a quarter of a mile plus a hundred yards.

Both Sallee and Rumsey record briefly the crew's meeting with Dodge and Harrison after those two had left the front of the fire. Sallee reports Dodge as saying that all of them "had better get out of that thick reproduction" because "it was a death trap" and then instructing Hellman to return the crew to the north side of the gulch and head them down the canyon to the river. Rumsey and Sallee agree Dodge didn't look particularly worried: "Dodge has a characteristic in him," Rumsey told the Board. "It is hard to tell what he is thinking." And Dodge probably wasn't yet alarmed, since he told Hellman that, while the crew was proceeding toward the river, he and Harrison would return to the cargo area at the head of the gulch and, as the others had already done, eat something before starting on the trail.

Still, it is clear Dodge hadn't cared for what he saw when he took a look at the front of the fire. He said it was not possible to get closer to the flames than one hundred feet and the "thick reproduction" he was worried about was a thicket of second-growth Ponderosa pine and Douglas fir that had sprung up after an earlier fire and was tightly interlaced and highly explosive, especially with the wind blowing upgulch. Primarily, the retreat to the river was for the safety of the crew, but if the wind continued to blow upgulch, the crew could attack the lower end of the fire from its rear or flanks to keep it from spreading, especially into Meriwether Canyon, which, like a good chimney, drew a strong updraft. If worse came to worst and the wind changed and blew downgulch, the crew could always escape into the river.

Dodge gave Hellman still another order—not to take the crew down the bottom of the gulch but to "follow contour" on the other slope, by which he meant that the crew should stay on the sidehill and keep on an elevation from which they could always see how the main fire on the opposite side was developing.

Hellman led the crew across the gulch and started angling for the river, and, sure enough, it happened as it nearly always does when the second-in-command takes charge. The crew got separated and confused—considering the short time Dodge was gone, highly confused and separated by quite a distance. Sallee says they ended up in two groups, five hundred feet apart, far enough apart that they couldn't see each other, and so confused that Sallee's group thought they were in the rear only to have to stop and wait for the rear group to catch up. Rumsey says that part of the time Navon, the former paratrooper from Bastogne, was in the lead. He was the one really professional jumper—and professional adventurer—among them and evidently was always something of his own boss and boss of the whole outfit if it looked to him as if it needed one.

This is all that happened in the twenty minutes Dodge was gone. But instead of being just a lunch break for the boss, it also was something of a prelude to the end. At least it can make us ask ahead of time what the structure of a small outfit should be when its business is to meet sudden danger and prevent disaster.

In the Smokejumpers the foreman is nearly always in the lead and the second-in-command is in the rear. On the march, the foreman sizes up the situation, makes the decision, yells back the orders, picks the trail, and sets the pace. The second-in-command repeats the orders, sees that they're understood, and sees that the crew is always acting as a crew, which means seeing that the crew is carrying out the boss's orders. When they hit a fire, the foreman again is out in front deciding where the fire-line should go and the second-in-command is again in the rear. He repeats the foreman's orders, he pats his men on the back or yells at them, and only if he can't himself get them to do what they should does he yell to the foreman, "They're making lousy line."

Although the foreman has little direct contact with his men, even on a friendly basis, his first job is to see that his men are safe. He is always asking himself, Where is a good escape route?

It is easy to forget about the second-in-command, who has a real tough job. He is the one who has to get the yardage out of the men, so he has to know how to pat them or yell at them and when. He has to know his men and up to a point be one of them, but he has to know where that point is. Being second in command, he will have a hard time, especially when he first takes command. A little friendship goes a long way when it comes to command, and they say Hellman was a wonderful fellow, but that may be part of the reason why, when he first took command, the outfit became separated and confused.

It could also have been partly the crew's fault. Now that they weren't going to hit the fire head on, some of the excitement was gone. Fighting a fire from its rear is not unusual, but it doesn't show how much horsepower you have. The crew, though, was still happy. They were not in that high state of bliss they had been in when they expected to have the fire out by tomorrow morning and possibly be home that same night to observe tall dames top-heavy with beer topple off bar stools. On the other hand, attacking the fire from the rear would make the job last longer and mean more money, and, in a Smokejumper's descending states of happiness, after women comes overtime. Actually, the priority could be the other way around. To the crew the fire was nothing to worry about.

Dodge felt otherwise as he and Harrison sat eating at the cargo area near the head of the canyon, from where he could see almost to the river. He told the Board of Review, "The fire had started to boil up, and I figured it was necessary to rejoin my crew and try to get out of the canyon as soon as possible."

He picked up a can of Irish white potatoes and caught up to his crew roughly twenty minutes after he had left them. It was "about 5:40," according to his testimony. Dodge had Hellman collect the crew, then station himself at the end of the line to keep it together this time while he himself took the lead and headed for the river. Things went fast from then on but never fast enough for the crew to catch up and keep ahead of disaster.

Rod Norum, who is one of the leading specialists on fire behavior in the Forest Service and still a fine athlete, as an experiment started out where Dodge rejoined his crew and, moving as fast as possible all the way, did not get to the grave markers as fast as the bodies did. Of course, there was nothing roaring behind him.

———

When the crew crossed back from the south to the north side of Mann Gulch where they had landed, they crossed from one geography into another and from one fire hazard into one they had never dealt with before. Mann Gulch is a composition in miniature of the spectacular change in topography that is pressed together by the Gates of the Mountains. Suddenly the Great Plains disappear; suddenly the vast Rocky Mountains begin. Between them, there is only a gulch or two like Mann Gulch for a transition from one world to another. Before the fire the two sides of Mann Gulch almost evenly divided the two topographical and fuel worlds between them—a side of the gulch for each world. The south side, where the fire had started, was heavily timbered. In the formal description of the *Report of Board of Review*: "At the point of origin of the fire the fuel type consisted of a dense stand of six- to eight-inch-diameter Douglas fir and some ponderosa pine on the lateral ridges."

But it was a different type of fuel on the north side, where the crew was now on its way to the river. "At the point of disaster the tree cover consisted of stringers of scattered young ponderosa pine trees with occasional overmature ponderosa pine trees. The ground cover or understory which predominated was bunch grass with some cheat grass." Essentially the north side of Mann Gulch was rocky and steep with a lot of grass and brush and only a scattering of trees. The south side was densely timbered.

The difference between the two sides of the gulch is after all these years still clearly visible. On the south side the charred trees stood until their roots rotted. Then winds blowing upgulch from the Missouri left them on the ground, unburied but paralleling each other, as if they belonged to some nature cult ultimately joined together by the belief that death lay in the same direction. At times they look as if they had been placed there—black-draped coffins from some vast battle awaiting burial in a national cemetery on a hillside near a great river, if not the Potomac then the Missouri.

On the north side, where the crew was angling toward the river, there are white crosses with bronze plates and a few black odds and ends. Not much else. The men died in dead grass on the north slope.

Several generalizations will help with what lies ahead if we remember that they are only generalizations. A fire in dense timber builds up terrific heat but not great speed. As Harry Gisborne has said, a big run for a crown fire is from half a mile to a mile an hour. A grass fire, by comparison, is usually a thin fire; it builds up no great wall of heat—it comes and is gone, sometimes so fast that the top of the grass is scarcely burned. Sometimes so fast it doesn't even stop to burn a homesteader's log cabin. It just burns over and around it, and doesn't take time to wait for the roof to catch on fire. Even so, since the great fire catastrophe of 1910, far more men had been killed by 1949 on fast, thin-fueled grass fires east of the Continental Divide in Montana than on the slow, powerful fires in the dense forests of western Montana.

Arthur D. Moir, Jr., supervisor of the Helena National Forest, generalizing in his 1949 testimony about the Mann Gulch fire, said fires east of the Continental Divide in Montana "are smaller, and because of less fuel, are more quickly controlled." But he went on to add that to his knowledge "only two men have been burned in forest fires in Idaho and western Montana since 1910," whereas he counted thirty-five who had burned to death east of the Divide in fast grass.

The grass and brush of Mann Gulch could not be faster than it was now. The year before the fire, the Gates of the Mountains had been designated a wilderness area, so no livestock grazed in Mann Gulch, with the result that the grass in places was waist high. Since it was early August with blistering heat, the worst of both fire-worlds could occur—if a fire started in the deep timber of the southern side, where most fires start, and then jumped to the explosive grass and shrubs of the northern side, as this one might, and did, it could burn with the speed of one of those catastrophic fires in the dry gulches of suburban Los Angeles but carry with it the heat of the 1910 timber fires of Montana and Idaho. It could run so fast you couldn't escape it and it could be so hot it could burn out your lungs before it caught you.

Things got faster and shorter. Dodge says they continued downgulch about five minutes, Sallee says between an eighth and a quarter of a mile, which is saying about the same thing. Dodge was worried—evidently no one else was. The fire was just across the gulch to be looked at, and that's evidently what they were doing. They were high enough up the slope now that they could almost peer into its insides. When the smoke would lift, they could see flames flapping fiercely back and forth, a damn bad sign but they found it interesting.

Navon was in his element as a freewheeler, alternating between being benevolent and being boss. He had lightened Rumsey's load by trading him his saw for Rumsey's heavier water can, so Rumsey especially was watching the scenery as it went by. He observed that the fire was burning "more fiercely" than before. "A very interesting spectacle," he told the Board of Review. "That was about all we thought about it."

Of the stations of the cross they were to pass, this was

the aesthetic one. On forest fires there are moments almost solely for beauty. Such moments are of short duration.

———

Then Dodge saw it. Rumsey and Sallee didn't, and probably none of the rest of the crew did either. Dodge was thirty-three and foreman and was supposed to see; he was in front where he could see. Besides, he hadn't liked what he had seen when he looked down the canyon after he and Harrison had returned to the landing area to get something to eat, so his seeing powers were doubly on the alert. Rumsey and Sallee were young and they were crew and were carrying tools and rubbernecking at the fire across the gulch. Dodge takes only a few words to say what the "it" was he saw next: "We continued down the canyon for approximately five minutes before I could see that the fire had crossed Mann Gulch and was coming up the ridge toward us."

Neither Rumsey nor Sallee could see the fire that was now on their side of the gulch, but both could see smoke coming toward them over a hogback directly in front. As for the main fire across the gulch, it still looked about the same to them, "confined to the upper third of the slope."

At the Review, Dodge estimated they had a 150- to 200-yard head start on the fire coming at them on the north side of the gulch. He immediately reversed direction and started back up the canyon, angling toward the top of the ridge on a steep grade. When asked why he didn't go straight for the top there and then, he answered that the ground was too rocky and steep and the fire was coming too fast to dare to go at right angles to it.

You may ask yourself how it was that of the crew only Rumsey and Sallee survived. If you had known ahead of time that only two would survive, you probably never would have picked these two—they were first-year jumpers, this was the first fire they had ever jumped on, Sallee was one year younger than the minimum age, and around the base they were known as roommates who had a pretty good time for themselves. They both became big operators in the world of the woods and prairies, and part of this story will be to find them and ask them why they think they alone survived, but even if ultimately your answer or theirs seems incomplete, this seems a good place to start asking the question. In their statements soon after the fire, both say that the moment Dodge reversed the route of the crew they became alarmed, for, even if they couldn't see the fire, Dodge's order was to run from one. They reacted in seconds or less. They had been traveling at the end of the line because they were carrying unsheathed saws. When the head of the line started its switchback, Rumsey and Sallee left their positions at the end of the line, put on extra speed, and headed straight uphill, connecting with the front of the line to drop into it right behind Dodge.

They were all traveling at top speed, all except Navon. He was stopping to take snapshots.

———

The world was getting faster, smaller, and louder, so much faster that for the first time there are random differences among the survivors about how far apart things were. Dodge says it wasn't until one thousand to fifteen hundred feet after the crew had changed directions that he gave the order for the heavy tools to be dropped. Sallee says it was only two hundred yards, and Rumsey can't remember. Whether they had traveled five hundred yards or two hundred yards, the new fire coming up the gulch toward them was coming faster than they had been going. Sallee says, "By the time we dropped our packs and tools the fire was probably not much over a hundred yards behind us, and it seemed to me that it was getting ahead of us both above and below." If the fire was only a hundred yards behind now, it had gained a lot of ground on them since they had reversed directions, and Rumsey says he could never remember going faster in his life than he had for the last five hundred yards.

Dodge testifies that this was the first time he had tried to communicate with his men since rejoining them at the head of the gulch, and he is reported as saying—for the second time—something about "getting out of this death trap." When asked by the Board of Review if he had explained to the men the danger they were in, he looked at the Board in amazement, as if the Board had never been outside the city limits and wouldn't know sawdust if they saw it in a pile. It was getting late for talk anyway. What could anybody hear? It roared from behind, below, and across, and the crew, inside it, was shut out from all but a small piece of the outside world.

They had come to the station of the cross where something you want to see and can't shuts out the sight of everything that otherwise could be seen. Rumsey says again and again what the something was he couldn't see. "The top of the ridge, the top of the ridge.

"I had noticed that a fire will wear out when it reaches the top of a ridge. I started putting on steam thinking if I could get to the top of the ridge I would be safe.

"I kept thinking the ridge—if I can make it. On the ridge I will be safe. . . . I forgot to mention I could not definitely see the ridge from where we were. We kept running up since it had to be there somewhere. Might be a mile and a half or a hundred feet—I had no idea."

The survivors say they weren't panicked, and something like that is probably true. Smokejumpers are selected for being tough, but Dodge's men were very young and, as he testified, none of them had been on a blowup before and they were getting exhausted and confused. The world roared at them—there was no safe place inside and there was almost no outside. By now they were short of breath from the exertion of their climbing and their lungs were being seared by the heat. A world was coming where no organ of the body had consciousness but the lungs.

Dodge's order was to throw away just their packs and heavy tools, but to his surprise some of them had already thrown away all their equipment. On the other hand, some of them wouldn't abandon their heavy tools, even after Dodge's order. Diettert, one of the most intelligent of the crew, continued carrying both his tools until Rumsey caught up with him, took his shovel, and leaned it against a pine tree. Just a little farther on, Rumsey and Sallee passed the recreation guard, Jim Harrison, who, having been on the fire all afternoon, was now exhausted. He was sitting with his heavy pack on and was making no effort to take it off, and Rumsey and Sallee wondered numbly why he didn't but no one stopped to suggest he get on his feet or gave him a hand to help him up. It was even too late to pray for him. Afterwards, his ranger wrote his mother and, struggling for something to say that would comfort her, told her that her son always attended mass when he could.

It was way over one hundred degrees. Except for some scattered timber, the slope was mostly hot rock slides and grass dried to hay.

It was becoming a world where thought that could be described as such was done largely by fixations. Thought consisted in repeating over and over something that had been said in a training course or at least by somebody older than you.

Critical distances shortened. It had been a quarter of a

mile from where Dodge had rejoined his crew to where he had the crew reverse direction. From there they had gone only five hundred yards at the most before he realized the fire was gaining on them so rapidly that the men should discard whatever was heavy.

The next station of the cross was only seventy-five yards ahead. There they came to the edge of scattered timber with a grassy slope ahead. There they could see what is really not possible to see: the center of a blowup. It is really not possible to see the center of a blowup because the smoke only occasionally lifts, and when it does all that can be seen are pieces, pieces of death flying around looking for you—burning cones, branches circling on wings, a log in flight without a propeller. Below in the bottom of the gulch was a great roar without visible flames but blown with winds on fire. Now, for the first time, they could have seen to the head of the gulch if they had been looking that way. And now, for the first time, to their left the top of the ridge was visible, looking when the smoke parted to be not more than two hundred yards away.

Navon had already left the line and on his own was angling for the top. Having been at Bastogne, he thought he had come to know the deepest of secrets—how death can be avoided—and, as if he did, he had put away his camera. But if he really knew at that moment how death could be avoided, he would have had to know the answers to two questions: How could fires be burning in all directions and be burning right at you? And how could those invisible and present only by a roar all be roaring at you?

———

On the open slope ahead of the timber Dodge was lighting a fire in the bunch grass with a "gofer" match. He was to say later at the Review that he did not think he or his crew could make the two hundred yards to the top of the ridge. He was also to estimate that the men had about thirty seconds before the fire would roar over them.

Dodge's fire did not disturb Rumsey's fixation. Speaking of Dodge lighting his own fire, Rumsey said, "I remember thinking that that was a very good idea, but I don't remember what I thought it was good for . . . . I kept thinking the ridge—if I can make it. On the ridge I will be safe."

Sallee was with Rumsey. Diettert, who before being called to the fire had been working on a project with Rumsey, was the third in the bunch that reached Dodge. On a summer day in 1978, twenty-nine years later, Sallee and I stood on what we thought was the same spot. Sallee said, "I saw him bend over and light a fire with a match. I thought, With the fire almost on our back, what the hell is the boss doing lighting another fire in front of us?"

It shouldn't be hard to imagine just what most of the crew must have thought when they first looked across the open hillside and saw their boss seemingly playing with a matchbook in dry grass. Although the Mann Gulch fire occurred early in the history of the Smokejumpers, it is still their special tragedy, the one in which their crew suffered almost a total loss and the only one in which their loss came from the fire itself. It is also the only fire any member of the Forest Service had ever seen or heard of in which the foreman got out ahead of his crew only to light a fire in advance of the fire he and his crew were trying to escape. In case I hadn't understood him the first time, Sallee repeated, "We thought he must have gone nuts." A few minutes later his fire became more spectacular still, when Sallee, having reached the top of the ridge, looked back and saw the foreman enter his own fire and lie down in its hot ashes to let the main fire pass over him.

4

When at approximately four o'clock that afternoon the parachute on the radio had failed to open, the world had been immediately reduced to a two-and-a-half-mile gulch, and of this small, steep world sixty acres had been occupied by fire. Now, a little less than two hours later, the world was drastically reduced from that—to the 150 yards between the Smokejumpers and the fire that in minutes would catch up to them, to the roar below them that was all there was left of the bottom of the gulch, and to the head of the gulch that at the moment was smoke about to roar.

Somewhere beyond thought, however, there was an outside world with some good men in it. There were a lot more men sitting in bars who were out of drinking money and also out of shape and had never been on a fire before they found themselves on this one. There are also times, especially as the world is blowing up, when even good men land at the mouth of the wrong gulch, forget to bring litters even though they are a rescue team, and, after having gone back to get some blankets, show up with only one for all those who would be cold that night from burns and suffering.

One good man from the outside had come close to the inside. At about five o'clock, roughly about the time the crew had tooled up and started from the head of the gulch toward the fire, one good man had started up Mann Gulch from the river, had seen the fire blow up around him, then had been trapped by it and lost consciousness while running through a whirl that had jumped to the north side of the gulch.

So about the time that Dodge and his crew were hurrying

downgulch into the blowup, ranger Robert Jansson at the lower end was running from the blowup back to the river.

Jansson was ranger of the Helena National Forest's Canyon Ferry District, its station located then on the Missouri River about twenty-three miles northeast of Helena. With roads now nearly everywhere, planes and helicopters overhead, and word spread almost instantly by telephone or radio, it has become much easier and quicker to get men on fires, and, as a result, many Forest Service districts have merged and disappeared. Today, Canyon Ferry District and the Helena District are one, but in 1949 the Canyon Ferry District alone was a big piece of country, somewhere between two hundred fifty thousand and three hundred thousand acres and extending far enough downriver to include Mann Gulch. Mann Gulch at the time of the fire was the responsibility of a dedicated ranger, not easy to work for, and, like others who came close to this fire and survived, he was never to escape it. As a ranger he had always been obsessive in his dedication; as often as he could, he walked on mountains, watching and working and expecting his men to do the same. When, for instance, he should have kept only a skeleton crew on the Fourth of July and let the rest of his crew go to town, he kept all his men at work chasing fires he had faked, and he was a Methodist and did not drink or smoke. Unlike most rangers, he was a sensitive, vivid writer, and both his official position and the things in him that saw and felt compelled him to write report after report on the Mann Gulch fire, two of which should be listed in the literature of forest fires, if there is such a listing and such a literature. His reports are in official prose and depend upon reporting and not prose, and so have to be exceptionally good reporting to be good.

Jansson starts fitting together when we come to understand his love of his district, his sense that it was his protectorate, and his constant fear that it was about to blow up or already had and he didn't know where. He was constantly in the field even when the district would have been better off if he had been in the ranger station. But he knew his district inch by inch and had made a fire plan for every gulch and every possible blowup, and he worked his men illegal hours and holidays so that the blowup would be averted or at least spotted immediately. He makes us think of ourselves as we would be if we were made responsible for two hundred fifty thousand or three hundred thousand acres of some of the roughest country anywhere on the Missouri River, and knew that, when the fire blew, our help would be mostly bums picked up in bars and college students on summer vacation. We would be trying to do everything ourselves all at once all the time and trying to get everybody else to do the same, and we would have been, as he was, chronically ill, telling his close friends that it was hereditary but never saying what "it" was.

There was a big streak in him of those old-time rangers who had 1910-on-the-brain. Rangers for decades after were on the watch for fear that 1910 might start again and right in their woodpile. Some even lost their jobs because a fire got away from them. So Jansson was part old-time ranger, part us, and part John Robert Jansson.

In fact, he had been one of the first to spot and report the Mann Gulch fire. About four o'clock the previous afternoon, August 4, a lightning storm had passed over Helena and the Gates of the Mountains while Jansson was in Helena attending a meeting with several other rangers in the office of supervisor Moir. Lightning was heavy and immediately eight fires were reported, four in Jansson's Canyon Ferry District. The rangers hurried off to get on their fires. Jansson spent the rest of that afternoon and the evening "taking suppression action on three lightning fires and one camper fire."

In addition to fires that appear almost immediately after a lightning storm, there are usually a few "sleepers" that take a day or more before they can be spotted, fires often caused by lightning striking a dead tree with only rocks and a few dead needles underneath, so that only after a time do enough ashes fall to get a ground fire started with smoke showing. Accordingly, Jansson and his supervisor agreed that Jansson should wait until eleven the next morning before making an air patrol in order to give the dew time to dry and the sleepers to awake.

On August 4, the day the lightning storm blew by, fire conditions were not critical. On August 5 they were. At Canyon Ferry Ranger Station, the fire danger rating was 16 out of a possible 100 on August 4, but on the next day it was 74 out of 100; when questioned as to how such a rating should be classified, Jansson answered, "Explosive stage."

Jansson had tried to reach Jim Harrison, the guard at Meriwether Station, on the evening of August 4 by radio, but atmospheric conditions were bad and it was the next morning before he made contact. Harrison told him that there had been many lightning strikes north (downriver in the direction of Mann Gulch) the afternoon before, but Jansson told him he should clean up the picnic grounds first and not start on patrol until eleven o'clock so he would miss no late sleepers.

The next morning at eleven o'clock Jansson flew from Helena down the Missouri to within a mile west of where the Mann Gulch fire was to appear and on the way back flew directly over the spot. But he returned to Helena having seen only one small smoke coming from a sleeper reported earlier in the morning. He even flew low over Harrison's patrol station, hoping to get a signal from him, but Harrison by now had probably spotted the fire and was on his way back to his Meriwether camp, where, after trying unsuccessfully to radio Canyon Ferry Ranger Station, he tacked his last message on his door. He and his mother did not know it, but he was on his way to rejoin the jumpers, and they were on their way to all that she feared.

Although nothing new had shown up from the air, the moment Jansson's plane landed at Helena a smoke rose from the Missouri but seemingly farther downriver than they had flown. When Jansson reported this new smoke to the supervisor's office, he was told that the lookout on Colorado Mountain had just reported a fire in Mann Gulch, but Jansson did not take this location as gospel since Colorado Mountain was about thirty miles from Mann Gulch. Not being sure that this new fire was even on Forest Service land, he decided to fly the Gates of the Mountains once more.

Now smoke could be seen from the river as his plane approached Mann Gulch. The fire was burning on the ridge between Mann and Meriwether but at that time entirely on the Mann Gulch side and seemingly not spreading except at its upgulch edge. He estimated it to be eight acres in extent (six acres by a later survey). When asked at the Review whether it was "an exceptional experience for a fire to be eight acres an hour after a ranger had not been able to see it," Jansson answered, "In my judgment, that was unusual."

While still observing the Mann Gulch fire from the air, Jansson had spotted another fire to the south, which was to be called the "York fire." York is on Trout Creek, which has a number of summer and even permanent homes on it, so the damage a fire in that location could do was considerably greater than what one in Mann Gulch could do. In fact, there were times during the coming day and night when it looked as if the best men and equipment should be sent to

Trout Creek. In addition, the first sleeper found that morning was now starting to smoke up. Maybe men and equipment should be sent to it. Evidently it was going to be one of those days when one of the biggest problems in facing danger was to figure out the biggest danger and not to have a change of mind too often or too late or too soon.

It is hard to know what to do with all the detail that rises out of a fire. It rises out of a fire as thick as smoke and threatens to blot out everything—some of it is true but doesn't make any difference, some is just plain wrong, and some doesn't even exist, except in your mind, as you slowly discover long afterwards. Some of it, though, is true—and makes all the difference. The first half of the art of firefighting is learning to recognize a real piece of fire when you see one and not letting your supervisors talk you out of it. Some fires are more this way than others and are good practice for real life.

On the return of Jansson's plane to Helena, cloud formations were noted which the pilot said meant high and variable winds. Forest fires need high and variable winds, and can make their own updrafts once they get started. The clouds were cumulus clouds, white, uneven, and puffy, caused by updrafts from variable heat conditions on the ground. In this story about Smokejumpers, clouds and winds are to be closely observed, even to the end and even if no one can tell for sure what details are going to make a difference.

———

By the time they landed again in Helena, Jansson had made a tentative "man-and-equipment order." For the Mann Gulch fire, he ordered "fifty men with equipment and overhead, two Pacific pumps with three thousand feet of hose, and fifty sack lunches." Soon he added twenty-five Smokejumpers to the order, because he believed the fire to be in an almost inaccessible area and the only chance to hold it to a small fire would be to get men on it in another few hours. Supervisor Moir agreed and telephoned the fire desk in Missoula, but was told that, although enough jumpers were available at the base, all planes except one were on mission, and the one plane, a C-47, could hold only sixteen jumpers with equipment. Jansson and Moir said no to sending a second plane later in the afternoon. That would probably be too late to get the jumpers on the fire before dark. This was at 1:44. At 2:30 the one plane left Missoula.

In the meantime, Jansson ate lunch in the office while he waited for the fifty-man emergency crew to be rounded up. Which brings up again the practice the Forest Service used to make of recruiting large crews of volunteer firefighters for emergencies. In those days the Helena National Forest Service thought it had an agreement with the Bureau of Reclamation and the Canyon Constructors to send men on call. But as it turned out, it was the Bureau's understanding that the plan was to operate only when its men were off shift, and at the time of the Mann Gulch fire practically none of them were. So when not enough Smokejumpers could be dropped on a threatening fire and there were other threatening fires to think about, the Forest Service soon would be combing the bars for stool-bums, and stool-bums, apart from not being much good on a fire, are not easy to get off their stools. Like the rest of us but perhaps more so, they would rather remain alive on a stool exposed to the possibility of being converted by the Salvation Army than be found dead on a fire-line with a Pulaski smoldering in their hands.

So after a while when nobody showed up, Jansson and alternate ranger Henry Hersey left the office and started working the bars, ending up with ten, not fifty, customers who thought they needed a breath of fresh air. The ones who were still drunk when they got to the fire may well have been better off than the ones who had sobered up and found out where they were.

Jansson located a Forest Service horse truck, loaded it with bedrolls, put the ten recruits on top of the rolls, and at 2:20 started for Hilger Landing twenty miles from Helena and nearly another six miles by boat from Mann Gulch. The owner of an excursion boat had telephoned to report the fire and had said he would hold his boat at Hilger Landing for the use of the Forest Service. Since it was 2:30 when the Smokejumpers left Missoula for Mann Gulch, Jansson and his crew were leaving Helena at about the same time, but it would be hours before they located each other.

By three o'clock Hersey had collected nine more revolving bar stools and started for Hilger Landing.

Jansson arrived at Hilger Landing only to discover that the owner of the excursion boat had left with a load of paying tourists so they could see firsthand the smoke coming out of the mouth of Mann Gulch. The ranger was sore as hell, but as a Methodist interested in working with young people he says only that he was "thoroughly peeved." There wasn't anything he could do, though, but wait. There was no road through the Gates of the Mountains in 1949, and because it is designated wilderness, the area still has no roads. Even today you need a boat to get through the cliffs of the Gates, and the boating is fine if you can get a boat, because some nine miles below the mouth of Mann Gulch there is a big power dam that has made a small lake of the river and backed it up clear to the upper end of the cliffs of the Gates, where Jansson was waiting with a crew and no paddle.

No wonder he was "peeved"; he had already worked out his plan with Hersey, and there was nothing constructive he could do while he waited. The plan was for both him and

Hersey to land at the mouth of Mann Gulch and then, since it would be getting late in the day and the crew was in poor shape, to set up camp there and wait until the next morning before hitting the fire. Hersey would take charge of the fire crew while Jansson would scout Mann Gulch, find the jumpers, and bring them back to camp, where they could put up a coordinated fight. It was a good plan, except that it did not allow for the wit of the universe and the mental lapses of man.

While Jansson waited at Hilger Landing, the recreation guard, Jim Harrison, was alone on the Mann Gulch fire. Jansson had to wait fifty minutes with his ten volunteers before the owner of a private boat, Fred Padbury, a Helena druggist, pulled up to the landing and took on Jansson and his crew and some of their equipment.

Soon after, the excursion boat returned to Hilger Landing and started down the river with Hersey, his nine drunks, and the remaining equipment. Of the total nineteen, only three had ever been on a fire before. Two were soon to be made straw bosses, an indication of what kind of fire crew they were going to be.

On the way down the Missouri now, Jansson came to a bend in the river where he could see that the fire had "slopped over" on the Meriwether side of the ridge. The "slop over" was already two or three acres, and the ranger was alarmed. The moment he saw the fire starting down the Meriwether slope he had to change plans—all would now land at Meriwether and Hersey would take the crew up the Mann Gulch–Meriwether trail to the top of the ridge and contain the fire on the Meriwether side. Then Jansson would continue downstream in the Padbury boat and scout the fire as originally planned. As usual, this plan left Jansson both the quarterback and the man in motion looking for trouble, and it's hard for even a good man to play two positions at once.

The excursion boat, being faster than the Padbury boat, passed Jansson and his crew before they reached the Meriwether landing, and Jansson was able to inform Hersey of his change in plans. But because the Padbury boat had to tread water while the big excursion boat docked and unloaded its equipment, Jansson had to wait some more. In the woods, as in the army, it's often a case of hurry up and wait, and Jansson did not get high marks in waiting.

The ranger told Hersey to take all nineteen volunteers and get on the fire as soon as possible, stopping only at the guard camp 150 yards up the canyon to radio an order to the supervisor's office. Hersey says he was to tell the office "that this was no training fire," "to get in gear, and give us all the support possible." Specifically, he was to order two new crews of fifty men each with "experienced overhead," one for Meriwether and one for Mann Gulch. He was also to tell Canyon Ferry Ranger Station to try to get in touch with the Smokejumpers by radio and to tell them if they were still in Mann Gulch to come down the Mann Gulch–Meriwether ridge to the slop over and join Hersey's nineteen men there. Then, before starting up the trail that comes out of Meriwether Canyon like a ladder about to topple backwards, Hersey gave his bar stools a lecture. He told them when they got to the fire they would try to hold two sides of it—the upgulch side so that the jumpers could get through and join them and the side approaching the Meriwether trail to keep their own escape route open and to save the tourist beauty of Meriwether Canyon.

They probably didn't get to the fire until nearly six o'clock. Jansson thought they should have been there sooner.

Jansson had left at about 4:35 in the Padbury boat to

carry out his plan to scout the fire and, if he found the jumpers, to bring them back to Meriwether. When the boat passed
the mouth of Mann Gulch, the fire was still on the upper
third of Mann Gulch's north-facing slope, although heavy
smoke was blowing from it across the gulch to the other
side. The Padbury boat continued on downstream until it
approached the mouth of Elkhorn Creek, a little more than a
mile below and to the northwest of Mann Gulch. As far away
as Elkhorn the air was feathered with ashes. The Padbury
boat then returned upstream to the mouth of Mann Gulch,
which by now was so full of smoke Jansson could not see up
it. If he wanted to know what was happening there, he would
have to walk. As usual, he was precise with figures. He says
he started up the gulch at 5:02.

At about that time Dodge's crew had collected the cargo
parachutes after their jump, had had something to eat, and
then had tooled up and started for the fire. So, at approximately the same time, the fire was being approached from
both its upper and its lower ends.

———

Jansson walked up the bottom of Mann Gulch for almost
half a mile, noting that the fire was picking up momentum
and still throwing smoke over his head to the north side of
the gulch where farther up Dodge had rejoined his crew and
was now leading them toward the river. Then right behind
Jansson at the bottom of the gulch a spot fire flowered. Then
several more flowered just below the main fire. Then a few
tossed themselves as bouquets across the gulch, grew rapidly
into each other's flames, and became a garden of wildfire.

What the ranger was about to see was the beginning of
the blowup. Seemingly without relation to reality or to the
workings of the imagination, the flowers that had grown

into a garden distended themselves into an enormous light bulb and a great mixed metaphor. Flowers and light bulbs don't seem to mix, but the light bulb of the mind strung itself inside with filaments of flame and flowers, bloated and rounded itself at its top with gases, then swirled upgulch to meet the Smokejumpers trying to escape downgulch. In a few minutes they met. Then only a few minutes later the blowup passed out of the gulch, blew its fuse, and left a world that is still burned out.

———

Jansson was probably the first to walk through a blowup propelled by a fire whirl, to drop unconscious in its vortex, to revive only a few feet from its flame, and to live to record it. Afterwards, he would keep returning to Mann Gulch with tape and stopwatch to check his original recordings of distances and times, and once he returned with objective observers to check him. He struggled to determine whether what had happened really had and really had measurable boundaries on earth. But we should already know enough about him to expect that we are about to observe a rare phenomenon of nature through the eyes of an especially fine observer. Jansson was a good enough observer to have been picked in the early years of the serious study of fire behavior by Harry Gisborne as one of his select group of rangers to field-test some of the early theories of the Priest River Experiment Station. To Jansson, Gisborne was an idol, as he was to nearly all those who first approached the study of forest fires scientifically, and as he is to some of us still living.

Later in this story of the Mann Gulch fire, it won't be enough to follow along merely as observers of Jansson's observations. There are things we see now that he would not have noticed, and there are things he saw but couldn't explain that we can put together. We know now pretty much what happened, because in part it happened in Mann Gulch and made its own contribution to the greatly enlarged inventory of our present knowledge of forest fires and so of our knowledge of many things about the woods. It did not do this immediately, and it was far from the only fatal forest fire that led to the congressional appropriations for the three great National Forest Fire Laboratories, one with its two wind tunnels right beside the Smokejumper base in Missoula. But before the Fire Labs, there had to be other fatal fires, other parents and communities sharing the same grief, and other newspapers making public records of grief. There also had to come a time when some of the most important members of Congress were from logging states and were members of just the right committees. So it came finally not in God's time but in the considerably slower time of bureaucracies, yet it came.

Even to me. It's different with me now from when I first started climbing Mann Gulch. Now I carry inside me part of the purgation of its tragedy. It is the part of me and the tragedy that knows more about forests and fires because of this forest fire. If now the dead of this fire should awaken and I should be stopped beside a cross, I would no longer be nervous if asked the first and last question of life, How did it happen?

———

Although it was not until the 1950s that the science of fire behavior became sufficiently advanced to explain blowups, it will be helpful to store up here for later use what so fine an observer as Jansson saw as he drew near to this tragedy of winds and fire.

Jansson noted strange sights as he went up the gulch. For one thing, since he could see for only two hundred yards because of the dense smoke, he knew immediately something big was on its way. He could see and hear rocks rolling, displaced by the heat. He could see and hear dead snags break off, without causal explanation, and he could see and hear that "flames were beginning to whirl and roar." At first these flames just flapped back and forth, signs of unstable air. But the unstable air started to spiral and the flames began to swirl like little dust devils. Soon, however, they united to become something like a tornado, caused by fire and causing fire and perfectly named a "fire whirl." What he refers to as a "holocaust" is a still later development, one that occurred when these fire whirls were starting other fire whirls that were starting still other fire whirls. Beside him, around him, and in front of him was a vast uproar trying to break the sound barrier. Behind him, sounds were tapering off and becoming silent, as sounds were turning into lights. The world behind him was becoming a circle of lights about to be turned out. Hell may have such illumination preceding such blackness.

Fire whirls both intensify existing fire and cause new fires. Their rotating action is that of a giant vortex, and, as giants, they can reach two thousand degrees in temperature. Fires that become giants are giant smoke rings with a downdraft in the center which is full of deadly gases and, what is more deadly still, heat so great it has burned out much of the oxygen; the outer ring is an updraft sometimes reaching the edge of the atmosphere.

Some fire whirls, not all of them, are flame throwers. Some pick up burning cones and branches. Some of the giants pick up burning logs and toss them ahead, starting spot fires sometimes a long way ahead. When these spot fires unite, firefighters can be trapped between two fires, as Jansson was soon to be.

A crown fire, as we know, is racing if it advances a mile an hour, but a fire whirl can go almost as fast as the wind.

———

Although Jansson thought he had put out of mind the possibility that the jumpers or anything human besides himself could be in Mann Gulch, he began to hear metallic noises that sounded like men working. That's the sound of flames heard by those alive after the flames go by. It is the thinking of those living who think they can hear dead men still at work.

Even with the flames closing in, Jansson had to follow the sounds in his head another eighth of a mile upgulch until any possibility that men were working in Mann Gulch had been obliterated. While he walked that eighth of a mile, the crown fire on the north-facing slope behind him had burned to the bottom of the gulch and the spot fires that had jumped to the opposite slope had converged behind him into one fire coming upgulch at him.

This fire front on the south-facing slope was in a few minutes to become the blaze only seventy-five yards behind the Smokejumpers after they dumped their heavy tools to run faster, and the crown fire which Jansson saw moving into the bottom of the gulch was to become the roar the Smokejumpers heard below them at almost the same time. At almost the same time everything was closing in on them and Jansson.

At 5:30 Jansson turned back and started to get out of there quick but still walking. Then the fire began to whirl continuously. When a streamer from it swept by, he realized after a couple of whiffs that the whirl could "cook out his lungs." He began to run. Now, he says, the whirl "was practically upright. My position was in the vortex, which was rapidly narrowing. I held my breath as I crossed the wall.

There was no flame, just superheated air and gases and a lot of reflected heat from the crown fire. I conked out from a lack of oxygen, fell on my left elbow, causing a bursitis which later caused my arm to swell."

When he came to, "the black creep of the fire" was only a few feet behind him. He had fallen victim for a few seconds to the two major enemies that threaten fighters of big fires—toxic gases, especially carbon monoxide, and lack of oxygen from overexertion and from hot air burning out the oxygen.

When he finally reached the boat, at 5:41, he placed himself in the bow next to Mrs. Padbury and watched the whirl for a few minutes. He thought again about the sound of men working that he imagined he had heard and again put it out of mind. Then he smelled his own vomit, apologized to Mrs. Padbury, and moved off to the side.

At the Board of Review, he was asked this question, to which he gave a short answer:

GUSTAFSON: In looking back up Mann Gulch draw . . . what was the picture as to the fire at that time?
JANSSON: A blowup.

Before the Padbury boat reached Meriwether Landing, superintendent Moir was in midriver in a speedboat preparing to go downstream. When the two boats met, Jansson transferred to the speedboat, and they landed where they could climb to an observation point giving them a complete view of Mann Gulch. Jansson says, "At that time it was apparent that all of Mann Gulch had burned out, but it appeared that the big blowup in Mann Gulch was over."

"At that time" can only be estimated, but it was shortly after six o'clock. When all is said and done, we still accept the hands of Jim Harrison's watch, which were melted permanently at about four minutes to six, as marking approximately the time that the fire was catching up to the crew.

About twenty minutes had passed between the time that Jansson left the mouth of Mann Gulch and the time he turned to view the whole of it. Near at hand, trees still exploded from the heat of their own resin, at a distance vast sounds were being converted into silent lights and the lights were being turned off, and nowhere were there any longer noises as of men working.

For a time the Mann Gulch fire was to become extinct in the minds of the world outside Mann Gulch. Jansson and Moir returned upriver to the Meriwether camp, planning the next day. The newly discovered York fire took on even greater importance after Jansson heard someone on the York fire radio calling frantically for additional help. When Jansson was unable to get through on the radio to ascertain the facts, the assistant supervisor, Favre Eaton, was sent from Meriwether to take charge of the York fire. Too many fires were going in too many directions for anyone to think of the Smokejumpers. Jansson says that "during the three-way conference between Moir, Eaton, and myself at Meriwether I am positive that there was nothing said about the Smokejumpers." For some time the Smokejumpers even passed out of existence on radio, no one noticing that their whereabouts were unknown, everyone assuming that Smokejumpers were infallible firefighters and were taking care of themselves wherever they were. For some time after Eaton left for the York fire and the superintendent with him, Jansson's main concern was why his crew of drunks hadn't returned from the cliffs before it got dark and they fell off.

5

In this story of the outside world and the inside world with a fire between, the outside world of little screwups recedes now for a few hours to be taken over by the inside world of blowups, this time by a colossal blowup but shaped by little screwups that fitted together tighter and tighter until all became one and the same thing—the fateful blowup. Such is much of tragedy in modern times and probably always has been except that past tragedy refrained from speaking of its association with screwups and blowups.

This story some time ago left the inside world at its very center—Dodge had come out of the timber ahead of his crew, with the fire just behind. He saw that in front was high dry grass that would burn very fast, saw for the first time the top of the ridge at what he judged to be about two hundred yards above, put two and two together and decided that he and his crew couldn't make the two hundred yards, and almost instantly invented what was to become known as the "escape fire" by lighting a patch of bunch grass with a gofer match. In so doing, he started an argument that would remain hot long after the fire.

At the time it probably made no sense to anyone but Dodge to light a fire right in front of the main fire. It couldn't act as a backfire; there wasn't any time to run a fire-line along its upgulch edge to prevent it from being just an advance arm of the main fire. Uncontrolled, instead of being a backfire it might act as a spot fire on its way upgulch and bring fire from behind that much closer and sooner to the crew.

Dodge was starting to light a second fire with a second

match when he looked up and saw that his first fire had already burned one hundred square feet of grass up the slope. "This way," he kept calling to his crew behind. "This way." Many of the crew, as they came in sight of him, must have asked themselves, What's this dumb bastard doing? The smoke lifted twice so that everyone had a good chance to ask the question.

The crew must have been stretched nearly all the way from the edge of the timber to the center of the grassy clearing ahead, where Dodge was lighting his fire. Rumsey and Sallee say that the men did not panic, but by now all began to fear death and were in a race with it. The line had already assumed that erratic spread customary in a race where everything is at stake. When it comes to racing with death, all men are not created equal.

At the edge of the timber the crew for the first time could have seen to the head of the gulch where the fire, having moved up the south side of the gulch, was now circling. From the open clearing they also could see partway toward the bottom of the gulch, where it was presumably rocks that were exploding in smoke. They didn't have to look behind—they could feel the heat going to their lungs straight through their backs. From the edge of the clearing they could also see the top of the ridge for the first time. It wasn't one and a half miles away; to them it seemed only two hundred yards or so away. Why was this son of a bitch stopping to light another fire?

For the first time they could also see a reef twelve to twenty feet high running parallel to the top of the ridge and thirty yards or so below it. This piece of ancient ocean bottom keeps the top of the ridge from eroding, as the rock lid on the top of a butte on the plains keeps the butte from eroding into plains. But no one was thinking of geology or probably even of whether it would be hard to climb over, through, or around. At this moment, its only significance was that it seemed about two hundred yards away.

When the line reached its greatest extension, Rumsey and Sallee were at the head of it—they were the first to reach Dodge and his fire. Diettert was just behind them, and perhaps Hellman, although these two stand there separately forever and ask the same question, What did Rumsey and Sallee do right that we did wrong? For one thing, they stuck together; Diettert and Hellman went their separate ways.

The smoke will never roll away and leave a clear picture of the head of the line reaching Dodge and his burned bunch grass. Dodge later pictured the crew as strung out about 150 feet with at least eight men close enough together and close enough to him so that he could try to explain to them—but without stopping them—that they could not survive unless they got into his grass fire. At the Review, he made very clear that he believed there was not enough time left for them to make it to the top of the hill, and events came close to supporting his belief. In the roar and smoke he kept "hollering" at them—he was sure that at least those closest to him heard him and that those behind understood him from his actions. In smoke that swirled and made sounds, there was a pause, then somebody said, "To hell with that, I'm getting out of here," and a line of them followed the voice.

The line all headed in the same direction, but in the smoke Dodge could not see whether any of them looked back at him. He estimated that the main fire would hit them in thirty seconds.

In the smoke and roar Rumsey and Sallee saw a considerably different arrangement of characters and events from Dodge's. Indeed, even the roommates differ from each other. Both agree with Dodge, however, that the line was stretched out, with a group at the head close to Dodge, then a gap, and then the rest scattered over a distance that neither could estimate exactly but guessed to be nearly a hundred yards. In fact, when in the summer of 1978 Rumsey, Sallee, Laird Robinson, and I spent a day together in Mann Gulch, the two survivors told Laird and me they were now sure some of the crew had fallen so far behind that they were never close enough to Dodge to hear whatever he was saying. The implication of Dodge's account is that they all passed him by, but Rumsey and Sallee believed that some of them hadn't. As to the head of the column, Sallee limits it to three—himself and Rumsey plus Diettert, who was also a pal and had been working on the same project with Rumsey before the two of them were called to the fire. To these three, Rumsey adds Hellman, the second-in-command, and indeed suggests, with Dodge agreeing, that it was Hellman who said, "To hell with that, I'm getting out of here," and so furnishes the basis for the charge that Hellman was doubly guilty of insubordination by being near the head of the line after Dodge had ordered him to the rear and by encouraging the crew to ignore Dodge's order to remain with him and enter his fire. Rumsey's testimony, however, will never settle Hellman's place in the line and hence his role in the tragedy, for Sallee was positive and still is that Hellman was at the head of the line when Dodge ordered the men to drop their tools but that he then returned to the tail of it, repeating Dodge's order and remaining there to enforce it. So direct testimony leaves us with opposite opinions of Hellman's closing acts as second-in-command of Smokejumpers on their most tragic mission. Either he countermanded his superior and contributed to the tragedy or, according to Sallee, being the ideal second-in-command, he returned to the rear to see that all the crew carried out the foreman's orders and to keep their line intact.

An outline of the events that were immediately to come probably would not agree exactly with the testimony of any one of the survivors or make a composite of their testimony, as might be expected, but would be more like what follows, and even what follows will leave some of the most tragic events in mystery and litigation.

Rumsey, Sallee, and Diettert left Dodge as one group and took the same route to the reef; two of them survived. Some of the crew never got as high up the slope as Dodge's fire. Hellman reached the top of the ridge by another route and did not survive. The rest scattered over the hillside upgulch from the route taken by the first three, and none of those who scattered reached the top. As Sallee said the summer we were together in Mann Gulch, "No one could live who left Dodge even seconds after we did."

In fact, the testimony makes clear that Diettert, Rumsey, and Sallee scarcely stopped to listen to Dodge. As Rumsey says, "I was thinking only of my hide." He and Diettert turned and made for the top of the ridge. Sallee paused for only a moment, because he soon caught up with Diettert and Rumsey, and actually was the first to work his way through the opening in the reef above. When asked at the Review whether others of the crew were piling up behind while he stood watching Dodge light his fire, Sallee said, "I didn't notice, but I don't believe there were. Rumsey and Diettert went ahead—went on—I just hesitated for a minute and went on too."

In the roar of the main fire that was now only thirty seconds behind them they may not even have heard Dodge, and, if they did hear words, they couldn't have made out their meaning. Rumsey says, "I did not hear him say anything. There was a terrible roar from the main fire. Couldn't hear much."

It probably wasn't just the roar from without that precluded hearing. It was also the voice from inside Mount Sinai: "I kept thinking the ridge—if I can make it. On the ridge I will be safe. I went up the right-hand side of Dodge's fire."

Although Sallee stopped a moment for clarification, he also misunderstood Dodge's actions. "I understood that he wanted us to follow his fire up alongside and maybe that his fire would slow the other fire down." Like Rumsey, Sallee interpreted Dodge's fire as a buffer fire, set to burn straight up for the top and be a barrier between them and the main fire. And like Rumsey, Sallee followed the right edge of Dodge's fire to keep it between them and the fire that was coming up the gulch.

The question of how Hellman reached the top of the ridge after leaving Dodge at his fire cannot be answered with certainty. What is known is that he made his way from where Dodge lit his fire to the top of the ridge alone, that he was badly burned, that he joined up with Rumsey and Sallee after the main fire had passed, that he told Rumsey he had been burned at the top of the ridge, and that he died the next day in a hospital in Helena. The most convincing guess about how he reached the top of the ridge is Sallee's. When he and I stood on the ridge in the summer of 1978, I asked him about Hellman's route to the top and he said that naturally he had thought about it many times and was convinced there was only one explanation: while he, Rumsey, and Diettert followed the upgulch (right) side of Dodge's fire and so for important seconds at least used it as a buffer protecting them from the main fire coming upgulch, Hellman must have followed the opposite, or downgulch, side of Dodge's fire and so had no protection from the main fire, which caught him just before he could get over the ridge.

Sallee talks so often about everything happening in a

matter of seconds after he and Rumsey left Dodge's fire that at first it seems just a manner of speaking. But if you combine the known facts with your imagination and are a mountain climber and try to accompany Rumsey and Sallee to the top, you will know that to have lived you had to be young and tough and lucky.

And young and tough they were. In all weather Sallee had walked four country miles each way to school, and a lot of those eight miles he ran. He and Rumsey had been on tough projects all summer. They gave it everything they had, and everything was more, they said, than ever before or after.

As they approached the reef, its significance changed for the worse. They saw that the top of the ridge was beyond the reef, and unless they could find an opening in it, it would be the barrier keeping them from reaching the top. They might die in its detritus. The smoke lifted only twice, but they saw a crevice and steered by it even after it disappeared again. "There was an opening between large rocks, and I had my eye on that and I did not look either way," Sallee says.

Halfway up, the heat on Rumsey's back was so intense he forgot about Dodge's buffer fire, if that is what it was, and, having spotted the opening, headed straight for it. It was not only upslope but slightly upgulch and to the right. In the smoke nothing was important but this opening, which was like magnetic north—they could steer toward it when they couldn't see it. Rumsey was in the center. Sallee was even with him on his left; Diettert was just a few steps behind on his right.

The world compressed to a slit in the rocks. Rumsey and Sallee saw neither right nor left. When asked at the Review whether they saw pincers of fire closing in on them from the sides, they said no; they saw only straight ahead. Ahead they saw; behind they felt; they shut out the sides.

A smokejumper descends toward a forest fire in northwest Montana, 1969. Photograph by J. M. Greany, courtesy USDA Forest Service.

The Mann Gulch fire seen from the air, August, 1949.
Photograph courtesy USDA Forest Service.

Mann Gulch and the Missouri River, August 20, 1949.
Photograph courtesy USDA Forest Service.

Body retrieval in Mann Gulch, August 6, 1949.
Photograph by Dick Wilson, courtesy of the photographer.

Four-point buck which died with the Smokejumpers
lies on blackened slope after the fire. Photograph by Peter
Stackpole, *Life*, © 1949 Time Warner Inc.

Survivors Robert Sallee and Walter Rumsey after the Mann Gulch fire.
Photograph by Peter Stackpole / The LIFE Picture Collection / Getty Images.

LEGEND

1, 2, & 3   Lightning struck trees
4   Dodge met Harrison
X   Dodge ordered crew to north side of gulch*
Y   Dodge and Harrison rejoined crew; beginning of the crew's race*
5   Jansson turned back
6   Dodge and crew turned back
7   Dodge ordered heavy tools dropped
8   Dodge set escape fire
9   Dodge survived here
10   Rumsey and Sallee survived here
11   Jumping area (chutes assembled, burned)
12   Cargo assembly spot (burned)
13   Spot fires
14   Approximate fire perimeter at time of jumping and cargo dropping (3:10 P.M.–4:10 P.M.)
15   Helicopter landing spot

BODIES FOUND

A   Stanley J. Reba
B   Silas R. Thompson
C   Joseph B. Sylvia
D   James O. Harrison
E   Robert J. Bennett
F   Newton R. Thompson
G   Leonard L. Piper
H   Eldon E. Diettert
I   Marvin L. Sherman
J   David R. Navon
K   Philip R. McVey
L   Henry J. Thol, Jr.
M   William J. Hellman

*Points X and Y added by the author

"Part of Mann Gulch Fire Area" (1952 contour map). Department of Agriculture.

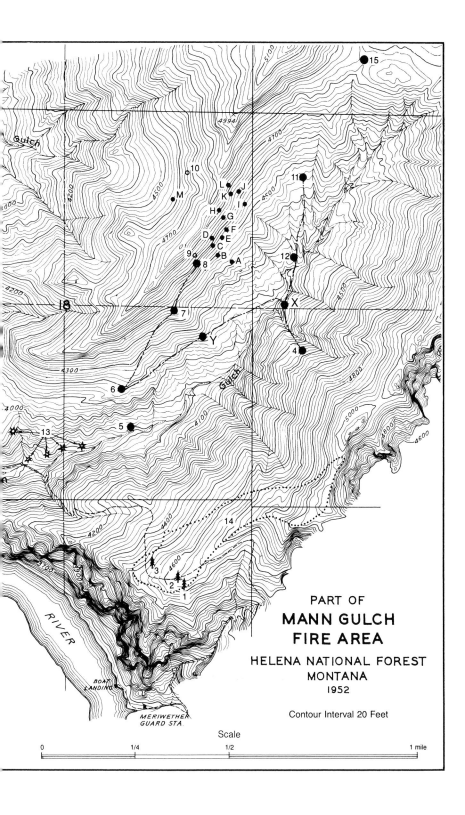

PART OF
**MANN GULCH
FIRE AREA**

HELENA NATIONAL FOREST
MONTANA
1952

Contour Interval 20 Feet

Scale

0   1/4   1/2   1 mile

Cross marking the site where Smokejumper Stanley J. Reba lost his life in the Mann Gulch fire. Photograph by Philip G. Schlamp, 1969. Courtesy USDA Forest Service.

Photo #5

+ lodge set Fire

Hellman recieved bu
before he crossed
over

Harri
Sylvia

Aug. 16, 1444

Taken f

Mann Gulch, August 16, 1949. Photography courtesy USDA Forest Service.

A Forest Service recreation guard on the Mann Gulch slope,
looking toward the Missouri River. Photograph by Philip G. Schlamp,
1969. Courtesy USDA Forest Service.

William Hellman's cross, Rescue Gulch. Photograph by Philip G. Schlamp, 1969. Courtesy USDA Forest Service.

Memorial plaque, Meriwether Campground. Photograph by Philip G. Schlamp, 1969. Courtesy USDA Forest Service.

To them the reef was another one of those things—
perhaps the final one—that kept coming out of smoke to
leave no place to run from death. They can remember feeling
sorry for themselves because they were so young. They also
tried not to think of anything they had done wrong for fear
it might appear in the flames. They thought God might have
made the opening and might take it away again. Besides, the
opening might be a trap for the sins of youth to venture into.

Beyond the opening and between it and the top of the
ridge they could see no flames but there was dense smoke.
Beyond the opening in the smoke there could be fire—
beyond, there could be more reefs, reefs without openings.
It could be that beyond the opening was the end of God and
the end of youth. Maybe that's what Diettert thought.

Rumsey and Sallee felt they were about to jump through
a door in a plane and so had to steady themselves and believe
something was out there that would hold them up. It was as
if there were a tap on the leg. Sallee was in the lead and was
first through the crevice. It was cooler, and he believed his
faith had been confirmed. He stopped to lower the temperature in his back and lungs. Rumsey was through next. As a
Methodist, he believed most deeply in what he had been first
taught. Early he had been taught that in a time of crisis the
top of a hill is safest. It was still some distance to the top, and
he never stopped till he got there.

Diettert stopped just short of the opening. On his birthday, not long after his birthday dinner and just short of the
top of the hill, he silently rejected the opening in the reef,
turned, and went upgulch parallel to the base of the reef,
where for some distance there is no other opening. No one
with him, neither Rumsey nor Sallee, saw him do this—it is
known by where his body was found. Diettert, the studious
one, had seen something in the opening he did not like, had

rejected it, and had gone looking for something he did not find. It is sometimes hard to understand fine students. Be sure, though, he had a theory, as fine students nearly always have.

While Sallee was cooling his lungs, he looked down and back at Dodge and the crew and for the first time realized why Dodge had lit his fire.

I saw Dodge jump over the burning edge of the fire he had set and saw him waving his arms and motioning for the other boys to follow him. At that instant I could see what I believe was all the balance of the crew. They were within twenty to fifty feet of Dodge and just outside the burning edge of the fire Dodge had set. The last I recall seeing the group of boys, they were angling up the slope in the unburned grass and fairly close to the burning edge of the fire Dodge had set . . . .

When Dodge first set the fire I did not understand that he wanted us boys to wait a few seconds and then get inside the burned-out grass area for protection from the main fire.

Dodge's description of his fire is mostly from inside it.

After walking around to the north side of the fire I started as an avenue of escape, I heard someone comment with these words, "To hell with that, I'm getting out of here!" and for all my hollering, I could not direct anyone into the burned area. I then walked through the flames toward the head of the fire into the inside and continued to holler at everyone who went by, but all failed to heed my instructions; and within seconds after the last man had passed, the main fire hit the area I was in.

When asked at the Review if any of the crew had looked his way as they went by, he said no, "They didn't seem to pay any attention. That is the part I didn't understand. They seemed to have something on their minds—all headed in one direction."

He wet his handkerchief from his canteen, put it over his mouth, and lay face down on the ground. Whether he knew it or not, there is usually some oxygen within fifteen inches of the ground, but even if he knew it, he needed a lot of luck besides oxygen to have lived, although Rumsey and Sallee were to say later that the whole crew would probably have survived if they had understood and followed Dodge's instructions.

It is doubtful, though, that the crew had the training and composure to interpret Dodge's instructions even if some of his words reached them over the roar. The close questioning Rumsey and Sallee received later at the Review revealed that their training in how to meet fire emergencies consisted of a small handful of instructions, four to be exact and only one of which had any bearing on their present emergency. The first was to backfire if they had time and the right situation, but they had neither. The second was to get to the top of the ridge where the fuel is usually thinner, where there are usually stretches of rock and shale, and where the winds usually meet and fluctuate. This is the one they tried, and it worked with only seconds to spare. The third instruction was designed to govern an emergency in which neither time nor situation permits backfiring or reaching a bare ridgetop. When it's that tough, the best you can do is turn into the fire and try to work through it, hoping to piece together burned-out stretches. The fourth and final warning was to remember that, whatever you do, you must not allow the fire to pick the spot where it hits you. The chances are it will hit you where it is burning fiercest and fastest. According to Dodge's later testimony, the fire about to hit them had a solid front 250 to 300 feet deep—no one works through that deep a front and lives.

Even if the crew's training had included a section on Dodge's escape fire, it is not certain that the crew would have listened to Dodge, would have entered the fire and buried their faces in the ashes. When asked at the Review if he would have gone into Dodge's fire had he received previous instruction about it, Rumsey replied, "I think that if I had seen it on a blackboard and seen it done and had it explained so that I understood it I think I surely would have gone in— but of course you never can tell for sure."

Dodge survived, and Rumsey and Sallee survived. Their means of survival differed. Rumsey and Sallee went for the top and relied on the soul and a fixation from basic training. The soul in a situation like this is mostly being young, in tune with time, and having good legs, an inflexible destination, and no paralyzing questions about what lies beyond the opening. When asked whether he had "ever been instructed in setting an escape fire," Dodge replied, "Not that I know of. It just seemed the logical thing to do." Being logical meant building one fire in front of another, lying down in its ashes, and breathing close to the ground on a slight elevation. He relied on logic of a kind and the others on time reduced to seconds. But no matter where you put your trust, at a time like this you have to be lucky.

The accounts that come down to us of the flight of the crew up the hillside nearly all conclude at this point, creating with detail only the happenings of those who survived, if only for a day, as Hellman did, or, like Diettert, at least reached the reef. Counting these two, only five are usually present in the story that goes on up the ridge. Only a sentence or two is given to those who, when last seen by Dodge, were all going in one direction and when seen finally by Sallee were angling through openings in the smoke below him as he looked down from the top of the ridge. Although they are the missing persons in this story, they are also its tragic victims. There is a simple aspect of historiography, of course, to explain why, after last seen by the living, they pass silently out of the story and their own tragedy until their tragedy is over and they are found as bodies: no one who lived saw their sufferings. The historian, for a variety of reasons, can limit his account to firsthand witnesses, although a shortage of firsthand witnesses probably does not explain completely why contemporary accounts of the Mann Gulch fire avert their eyes from the tragedy. If a storyteller thinks enough of storytelling to regard it as a calling, unlike a historian he cannot turn from the sufferings of his characters. A storyteller, unlike a historian, must follow compassion wherever it leads him. He must be able to accompany his characters, even into smoke and fire, and bear witness to what they thought and felt even when they themselves no longer knew. This story of the Mann Gulch fire will not end until it feels able to walk the final distance to the crosses with those who for the time being are blotted out by smoke. They were young and did not leave much behind them and need someone to remember them.

———

The Foreman, Dodge, also must be remembered, as well as his crew, and it is again the storyteller's special obligation to see that he is. History will determine the direction or directions in which the storyteller must look for his enduring memories, and history says Dodge must live or die in his escape fire. Ordinary history says he lived by lying in the ashes of his escape fire until the main fire swept over him and cooled enough to let him stand up and brush himself off. The controversial history that was soon to follow and has lasted ever since charges that Dodge's escape fire, set in front of the main fire, was the fire that actually burned some of the crew and cut off others from escaping. Historical questions the storyteller must face, although in a place of his own choosing, but his most immediate question as he faces new material is always, Will anything strange or wonderful happen here? The rights and wrongs come later and likewise the scientific knowhow.

The most strange and wonderful thing on the hillside as the escape fire swept up it, shutting it out of sight in smoke and heat, is that a spot of it remained cool. The one cool spot was inside Dodge. It was the "characteristic in him" that Rumsey had referred to when Dodge returned from the head of the fire with Harrison and muttered something about a death trap. It was the "characteristic" he was best known by, the part of him that always kept cool and aloof and believed on principle in thinking to itself and keeping its thoughts to itself because thinking out loud only got him into trouble. It was this characteristic in him that had started him to lead the crew downgulch to safety, then didn't like what it saw ahead and turned the crew back upgulch trying to outrun the fire without his ever explaining his thoughts to the crew. His running but not his thinking stopped when he saw the top of the ridge, for he immediately thought his crew could not make the top and so he immediately set his escape fire. When he tried to explain it, it was too late—no one understood him; except for himself, they passed it by. Except to him, whom it saved, his escape fire has only one kind of value—the value of a thought of a fire foreman in time of emergency judged purely as thought. The immediate answer to the storyteller's question about the escape fire is yes, it was strange and wonderful that, in this moment of time when only a moment was left, Dodge's head worked.

To see how Dodge's life as a woodsman shaped his thoughts in an emergency and to follow his thoughts closely, one more tick must be added to the tock of his makeup. In an emergency he thought with his hands. He had an unusual mechanical skill that helped him think, that at least structured his thoughts. It was a woodsman's mechanical skill—he liked to work with rifles, fix equipment, build lean-to's or log cabins. He wasn't fancy, he was handy. And in fact that spring he had been excused from training with the Smokejumpers so that he could be maintenance man for the whole Smokejumper base—no doubt part of the cause of the tragedy he was about to face with a crew only three of whom he knew. The foreman, then, was facing this tragic emergency alone, withdrawn as he often was into his own thoughts, which were the thoughts men and women have who are wired together in such a way that their brains can't start moving without their hands moving at the same time.

The present question, then, in its purest form is, How many brains, how much guts, did it take in those fiery seconds to conceive of starting another fire and lying down in it? In its maximum form, the question would be, Did Dodge actually make an invention when 250 or 300 feet of solid flames were about to catch up to him?

Two of the Forest Service's greatest fire experts, W. R. ("Bud") Moore and Edward G. Heilman, Moore's successor as director of fire control and aviation management for the Forest Service's Region One, have told me they never heard of this kind of escape fire before Dodge's use of it, and their experience corresponds to my own, which, though limited to summers when I was young, goes back to 1918. Rumsey and Sallee say under oath that in 1949 nothing like it had been mentioned in their training course, and, as Rumsey adds, even if it had been explained to him and he had seen it work, it seemed crazy enough so that he wasn't sure he would have stepped into it if it had been for real.

A lot of questions about the woods can't be answered by staying all the time in the woods, and it also works the other way—a lot of deep inner questions get no answer unless you go for a walk in the woods. My colleague at the University of Chicago Robert Ferguson pointed out to me that James Fenimore Cooper had something like Dodge's fire burning in his favorite of his own novels, *The Prairie*, first published in 1827. Cooper's eastern readers are held in suspense throughout most of chapter 13 by the approach of a great prairie fire from which the old trapper rescues his party at the last moment by lighting a fire in advance of the main one and having it ready for human occupancy by the time the sheet of flames arrives. He stepped his party into the burned-off grass and moved them from side to side as the main fire struck.

Cooper's readers clearly were not expected to know of this device or there would be no justification for the prolonged suspense which the chapter is supposed to create, but the escape fire on the prairie is no literary invention.

Mavis Loscheider of the Department of Anthropology at the University of Missouri, an outstanding authority on the life of the Plains Indians, sent me evidence showing that something like this kind of fire was traditionally set by Plains Indians to escape from grass fires and that pioneers on the plains picked up the invention from the Indians.

In his second volume of *The American Fur Trade of the Far West,* Hiram M. Chittenden describes how the prairie escape fire worked in the early 1800s:

> The usual method of avoiding the danger of these [prairie] fires was to start one in the immediate vicinity of the person or company in peril. This fire, at first small and harmless, would

soon burn over an area large enough to form a safe asylum and when the sweeping cohorts of flame came bearing down upon the apparently doomed company, the mighty line would part as if by prearrangement and pass harmlessly by on either side.

There are still good grounds, however, to believe Dodge "invented" his escape fire. Why doubt his word before the Board of Review that he had never heard of such a thing before? Even if it was known to mountain men, it could not have been much used in timbered country, if for no other reason than that it would seldom work there. The heat of a timber fire is too intense, and the fire is too slow and prolonged and consumes too much oxygen to permit walking around in it. Chances are Dodge's fire wouldn't have worked (wouldn't even have been thought of) if Dodge had been caught on the other, timbered side of Mann Gulch where the fire started. Moreover, Dodge's escape fire differs in important ways from the escape fires used by Indians and pioneers. Dodge's fire was started so close to the main fire that it had no chance to burn a large "asylum" in which the refugee could duck and dodge the main fire. Not being able to duck and dodge and remain alive, Dodge lay down in the ashes, where the heat was least intense and where he was close enough to the ground to find some oxygen.

Of course, Dodge had a Smokejumper's knowledge that if you can't reach the top of the hill you should turn and try to work back through burned-out areas in the front of a fire. But with the flames of the fire front solid and a hundred yards deep he had to invent the notion that he could burn a hole in the fire. Perhaps, though, his biggest invention was not to burn a hole in the fire but to lie down in it. Perhaps all he could patent about his invention was the courage to lie down in his fire. Like a lot of inventions, it could be crazy and consume the inventor. His invention, taking as much guts as logic, suffered the immediate fate of many other inventions—it was thought to be crazy by those who first saw it. Somebody said, "To hell with that," and they kept going, most of them to their deaths.

Dodge later told Earl Cooley that, when the fire went over him, he was lifted off the ground two or three times.

"This lasted approximately five minutes," he concludes in his testimony, and you and I are left to guess what the "this" was like. His watch said 6:10 when he sat up. By that time, death had come to Mann Gulch.

Dodge himself was allowed to live a little over five years more, what then was thought to be about the maximum time one who had Hodgkin's disease could live. However, he would never jump again. His wife knew when he entered the hospital for the last time that he knew it was for the last time. Like many woodsmen, he always carried a jackknife with him in his pants pocket, always. She told me that when he entered the hospital for the last time he left his jackknife home on his bedroom table, so he and she knew.

———

When Rumsey and Sallee reached the crevice, the main fire had reached the bottom of Dodge's escape fire. They were ahead of the flames, or at least thought they were, but couldn't be sure because of the rolling and un-revealing smoke. Rumsey fell into what he thought was a juniper bush and would not have bothered to get up if Sallee hadn't stopped and coldly looked at him. In the summer of 1978, when Rumsey and I were where he thought the juniper bush must have been, he said to me, "I guess I would be dead if he hadn't stopped. Funny thing, though, he never said a word to me. He just stood there until I said it to myself, but I don't think he said anything. He made me say it." They ran upgulch on the top of the ridge for a hundred yards or so and staggered down the slope on the other side of the ridge. There they stumbled onto a rock slide "several hundred feet long and perhaps seventy-five feet wide." The dimensions hardly seem large enough, but there weren't any other rock slides around. Within five minutes, the fire, coming down from the top of the ridge, had reached them.

Although Rumsey says they were both "half hysterical," they were objective enough to see that the fire as it approached them was following the patterns of a fire coming over a ridge and starting down the other side. At the top of the ridge it burned slowly, veering back and forth in the way fires do as winds from opposite sides of a ridge meet each other. It flapped, sometimes it turned downhill toward them, and once it turned sideways and jumped a draw with a spot fire and, well started there, it jumped back again. Once below the fluctuations at the top of the ridge it settled down and burned straight toward them. It burned with such intensity that it created an updraft, sucking in its center so that it was now a front with two pincers. It hit the rock slide on two sides. Rumsey and Sallee, like the early prairie pioneers, tried to duck and dodge in their asylum, but there wasn't much room for running. Rumsey says the fuel was thinner near the top of the ridge. "The flames were only eight to ten feet high."

A form like a solidification of smoke stumbled out of the smoke ahead and died in the rocks. It was a four-point buck burned hairless except for the eyelashes.

After the fire passed the rock slide "it really started rolling" downhill, replacing trees with torches.

Soon they heard someone calling from far off, but it turned out to be "only thirty yards away." It was Bill Hellman. His shoes and pants were burned off, and his flesh hung in patches. When asked at the Review, "Did Hellman at that time seem to be suffering tremendously?" Sallee answered, "Yes." To the next question, "Did he make any statement to you?" Sallee's reply was, "He just said to tell his wife something, but I can't remember what it was."

They laid him on a long, flat rock to keep his burns out of the ashes. As Rumsey says, "There wasn't much else we could do," having thrown away all their first-aid supplies on their flight from the fire.

Suddenly, there was a shout and a form in the smoke. It was Dodge answering the shouting that had gone on between them and Hellman. He "didn't appear excited," but he "looked kind of—well, you might say, dumbfounded or shocked." His eyes were red from smoke and his clothes black with ashes. He obviously was not his fastidious self, but he still had a characteristic about him.

They didn't say much about anything, least of all about whether the missing were alive. Dodge, in coming over the hill, had seen one alive and couldn't remember his name except that it began with "S" (Joe Sylvia). When Dodge sat up in his own fire he heard someone "holler" faintly to the east and, after a long time, found him only 150 to 200 feet upgulch and, oddly, below him, perhaps 100 feet. He was badly burned and euphorically happy. Dodge moved him to the shelter of a big rock and cut the shoes off his swollen feet, but there was no use in Dodge leaving his only worldly gift with him, his can of Irish white potatoes, since Sylvia could not feed himself with the charred and useless remains of his hands. In the hours to come, he would be without water because he could not lift his canteen.

Evidently Dodge hadn't seen any others as he came up the hill or crossed to the other side, and, as he said at the

Review, by the time he reached Rumsey, Sallee, and Hellman he "didn't think any of [the others] were still alive."

Rumsey and Sallee had come to a more hopeful conclusion once the fire passed them by in the rock slide—after all, they had made it, and, besides, once they understood the intention of Dodge's fire, they believed it would work and assumed at least some of the crew behind them had understood Dodge's fire and crowded into it. But Dodge's arrival eliminated that possibility, so there was very little they dared to talk about. After a while Dodge and Sallee left Hellman in Rumsey's care and started back uphill through the ashes without saying just why. Since none had been saved with Dodge, the assumption now was that any survivors would have made it over the hill, as Rumsey and Sallee had, so their search was a short one. Besides, the heat was still so intense it soon drove them back. They didn't have to explain why they didn't have anything to say when they returned.

It was getting dark. Hellman already had drunk most of their water, even though it made him sick. He could see the glare of the Missouri a mile and a half below, and it inflamed his thirst, but he was not allowed to think of walking. He did revive enough to become talkative. It was here that he told Rumsey he had been burned at the top of the ridge, and it was partly on the basis of this remark that Sallee formed his assumption that Hellman had reached the top of the ridge by following the downgulch side of Dodge's fire and so had had no buffer between him and the main fire raging upgulch. Once burned, though, like a wounded deer, he had started downhill for water but had collapsed after a few hundred yards. He was told to lie still on the rock and keep talking to forget the pain. Rumsey stayed with him, and at dusk Dodge and Sallee started for the river, Dodge leaving with them his canteen of water and his can of Irish white potatoes.

Dodge and Sallee had a tough time getting down to the river. They had to go half a mile or more before they could find a weak spot in the fire front through which to work their way. They had no map or compass, and when they reached the river they went the wrong way.

6

For the next few hours, the Smokejumpers who had landed in Mann Gulch passed from human remembrance perhaps as completely as they ever will. There were only five known to be alive at that moment, two of these soon to die, one with a name that began with "S."

Although Hellman had made it over the top of the ridge, he was despairing and smelling of burned flesh and was praying with Rumsey, who had been left to take care of him. They both had let their church attendance lapse and could not remember their prayers, so in embarrassment they prayed silently. From their position near the top of the ridge they could see, when the smoke opened, reflections of the fire in the Missouri River below, and Hellman had to be told again and again that he could not run to the river and immerse himself.

Dodge had left his can of white potatoes with Hellman because Rumsey would be there to feed him, but instead of eating the potatoes Hellman drank the salt water in the can and further inflamed his thirst.

For Dodge and Sallee on their way to the river, it was a never-never land in the night and the smoke, and without a map or compass. Both were near exhaustion and shock when they reached the river, and going downstream, which was easier for the water, also seemed easier for them. A boat passed that did not see them, then turned and went back upstream, and on this slight evidence they turned around too. They didn't know much about the world anymore, not even whether it was up or down.

Among other things, there were eleven of the crew they didn't know anything about. The missing were probably in a world one hundred by three hundred yards—the world between the boy with a canteen of water and no hands to lift it and Hellman on the other side of the ridge, who was looking for forgotten prayers.

The two top men in the Helena National Forest, supervisor Moir and assistant supervisor Eaton, had left Meriwether in haste for the fire at York because they and Jansson had agreed it was probably more important than the Mann Gulch fire. They had left in special haste because they could not reestablish radio contact with the crew on the York fire to determine its extent and the psychological stability of the crew that was fighting it. Matters got no more composed after they left when Jansson found out that the receiver of the radio at York had been dropped and broken by a hysterical volunteer sobbing for help. Both men and equipment were breaking.

Dodge and Sallee had been going downriver to nowhere. At the same time, coming downriver from above were hordes of picnickers. Full of beer and the desire to be mistaken for firefighters, they landed at the Meriwether picnic grounds and crowded into the guard station to hear whatever they could get near enough to hear. Soon it became impossible for Jansson to tell the picnickers from his volunteer barflies, so he tried by radio to stop all except official boating on the river, but the radio operator at Canyon Ferry was off somewhere.

Jansson forever held himself guilty for not being concerned about the jumpers at this time, although it is hard to see the justice of his continual prosecution of himself. As everyone did who did not think of them, he assumed the jumpers were too good to be caught in a fire—they either had joined up with Hersey's crew on the Meriwether ridge or had escaped over the head of the gulch into Willow Creek or perhaps hadn't liked the looks of things from the very first and had not jumped at all.

Once Jansson did try to radio Missoula about the jumpers, but another frequency kept cutting him off. Then he went back to the job of trying to get some coherence in his camp. As he knew, there is no better way to do this than to start a training school—he tried to make a fire foreman out of one of the three men who had been on a fire before, and he tried to make a radio operator out of another volunteer, but his best luck was with two picnickers pretending to be firefighters whom he trained to be camp cooks for a crew that had now grown to thirty-five. A mystery of the universe is how it has managed to survive with so much volunteer help.

———

Jansson was also keeping an eye on the fire near the top of Meriwether ridge where he had sent alternate ranger Hersey and his crew of nineteen men with two hold-at-all-cost orders: (1) hold the trail from the east open so that the jumpers could come down the ridge and join them, and (2) above all, hold the perpendicular trail behind them open so that, if they had to, they could escape back to the camp and, if they had to, from there into the river. The fire now was definitely moving down the Meriwether slope.

As it darkened, Jansson began to see flames making movies of themselves on the faces of the cliffs fifteen hundred feet straight above him.

Hersey says that when some of the crew saw the cliffs reenact the fire they tried to jump off them.

In the wide world, Hersey was probably the only man

in whose mind the Smokejumpers were constantly present. Their absence was heightened by the fact that Hersey had followed Harrison's tracks on the trail to the top of the ridge and the front of the fire—his tracks were easy to follow because he had been using his Pulaski as a walking stick, and going up that stepladder trail he had relied on it as if he and his Pulaski were a cripple and a crutch. What worried Hersey most was that at the top of the ridge Harrison's tracks headed into some second growth that the fire was already burning. Hersey spent most of his time organizing his crew into a fire-line and giving them another speech about how to face danger. He gave them a speech about facing danger every time he walked around the head of the fire and every time the fire ran a reel of itself on a cliff. It would be interesting to know what he told them, because it seems to have worked fairly well. Anyway, his crew stayed on the line even after the trail up the ridge to the east had disappeared in flames. His crew, though, were drinking gallons of water more than seasoned firefighters would have, so he had to send one of them all the way down to Meriwether Station with a canvas sack for another load.

When Jansson saw the waterbuck in camp, he became alarmed. Because the Smokejumpers had become nonexistent in Jansson's mind, it was his own men fifteen hundred feet above his head who, he thought, were in danger. The returned waterbuck was a sure sign that Hersey intended to stay on the ridge and fight, and Jansson did not want him fighting fire after dark on the edge of fifteen-hundred-foot pinnacles with a bunch of drunks. Now, for the first time, he also became alarmed about the jumpers, who, the waterbuck was telling him, had not shown up on the Meriwether ridge. If they weren't with Hersey's crew, there were few places in the world that they could reach where they would be safe.

Jansson immediately ordered the radio at Canyon Ferry to get the radio at Missoula to use all frequencies to locate the whereabouts of the jumpers. When he was advised by Missoula that it was unable to establish contact with the jumpers on any frequency, he then asked for the exact location of their jumping area. "While they were giving me the exact spot," he says, "foreman Dodge and jumper Sallee walked into the guard cabin at Meriwether and Dodge reported that he had two injured men. This was at approximately 8:50 P.M."

The volunteers and the picnickers and the drunks crowded into the cabin. Jansson had to take Dodge outside and up the canyon to get any coherent information from him, but what did Dodge know that was coherent? He knew that back in what earlier in the day had been Mann Gulch were two badly burned men, one with a name Dodge did not remember, and one unburned man, Rumsey, watching the burned man with a name, Hellman. What else was in the amphitheater for sure was fear and the smell of overcooked flesh.

Jansson immediately ordered through the Canyon Ferry radio one doctor, two litters, blankets, and blood plasma. At ten o'clock Hersey came in with his terrified postdrunks, having kept them on the fire until they had been several times trapped by it. He told Jansson about Harrison's tracks, and, even more alarming, he told Jansson he had seen no jumpers or their tracks.

"We decided," Jansson says, "to consider the rescue work the No. 1 job and the fire the No. 2 job. I asked Hersey to look after the fire job while I went for the jumpers."

———

It is like that in the woods and even in the wide world generally—the rescue of men and women, alive or dead, comes first. Of course, some step on the gas and leave them lying on the pavement where they landed and some sneak off, like Egyptian bas-relief, with their profiles looking one way and their bodies going the other way. But most people think they can be of help, and some even seem born to rescue others, as poets think they are. The best of them goof, especially at first, because only a few have the opportunity to keep in practice. Then as they catch on again they become beautiful in performance if one can step back for a moment to look. Almost as beautiful as when, having completed their job of deposing death, they fade into complete anonymity. It was very hard, for instance, to rescue the names of those Jansson picked for his rescue team. Even though he must have regarded them as his best, they all made mistakes, especially at first. But they also support the statement that one of the finest things men and women do is rescue men and women, even when they know they are rescuing the dead. This statement takes into account the Egyptian bas-relief, the drunks, and the sobbing radios.

———

At 10:30, while the rescuers were still waiting at Meriwether for the doctor and medical supplies to arrive, rumors and uncertainties were spreading through the camp. They spread in waves and, like waves, spent themselves draining into the sand, but one kept resurfacing—the rumor that there were injured men downriver waiting to be picked up. Jansson left in a speedboat hoping to bring back the eleven missing men, but the report turned out to be about Dodge and Sallee, who had been seen walking upriver by several boatloads of picnickers. This is a common enough way to start off a rescue operation—running after a rumor that turns out to be a misinterpretation of something already known.

For a while Jansson patrolled the lower river, signaling

with a flashlight and occasionally cutting off his motor and yelling. Finally a speedboat arrived with two doctors in it, T. L. Hawkins of Helena and his guest, R. E. Haines of Phoenix, Arizona, and Jansson transferred to their boat and landed at the mouth of Mann Gulch. Soon the big excursion boat with the rescue party in it arrived, only to discover they all were at the mouth of the wrong gulch—Dodge and Sallee had come down a gulch below Mann Gulch. When they arrived at this lower one that came to be called Rescue Gulch, they discovered that the litters had been left six miles back at Hilger Landing. Almost as soon as the speedboat started back to get them, rumors and tension mounted among the crew. One of the worst things a rescue crew does is wait— they wanted to start uphill immediately to find the injured men and let the litter crew come when the litters arrived. Jansson knew he had only one man who could lead them back through night and fire and rolling rocks and exploding trees—Sallee, who alone knew that he was just seventeen years old. Acting again on the assumption that the one sure way to quiet a crew is to get them to do something, Jansson lined them up and conducted roll call, only to discover he had six or seven men too many. They were picnickers who had smuggled themselves into the big excursion boat in the hope they could join the rescue crew. He had to cut them out and send them back. That left him with a crew of twelve, counting himself, the doctors, and Sallee, all tough men who had worked all day and now would work all night and probably the coming day in the agonizing valley.

It was 11:30 before Jansson and his crew started up Rescue Gulch. They had two litters but only one blanket, which, as it turned out, was all that was sent back to them when they had sent out for blankets. By now the insanity of the fire had passed on, and it lay twitching around its edges, like something dead but still with nerve ends. Its self-inflicted injuries had been great and had turned black. It lay in burned grass and split rocks with its passion spent. The crew crossed through the weakened fire-line into the world that might be dead.

About two-thirds of the way to the top they heard a shout, which turned out to be Rumsey coming down the hill to refill the canteen for Hellman, who had been drinking water, getting sick at the stomach, then drinking more water until he drank it all. Rumsey told Jansson that he thought his guard Harrison was dead, because when last seen Harrison had been sitting with his pack on his back not able to take it off. Rumsey didn't know if the others had survived.

Later at the Review, when Jansson was asked if Rumsey had made any detailed comment to him at this time about himself, Jansson replied, "He made the following comment, 'The Lord was good to me—he put wings on my feet and I ran like hell.'" This was one good Methodist talking to another.

Nearly half a mile away the crew could hear Hellman shouting for water. In the valley of ashes there was another sound—the occasional explosion of a dead tree that would blow to pieces when its resin became so hot it passed the point of ignition. There was little left alive to be frightened by the explosions. The rattlesnakes were dead or swimming the Missouri. The deer were also dead or swimming or euphoric. Mice and moles came out of their holes and, forgetting where their holes were, ran into the fire. Following the explosion that sent the moles and ashes running, a tree burst into flames that almost immediately died. Then the ashes settled down again to rest until they rose in clouds when the crew passed by.

Jansson, Rumsey, and Sallee pushed ahead of the main party to get water to Hellman. Jansson was the leader of the rescue crew, and he should tell it: "Hellman's face, arms, legs, and back were severely burned with loose flesh hanging in patches. He complained of the cold and was very thirsty. We let him rinse out his mouth and take on a little water. Water upset his stomach at first."

In ten or fifteen minutes the two doctors arrived. They gave Hellman a hypo and one quart of plasma, applied salve, transferred him to a litter, and then covered him with the one blanket. According to Jansson, "Bill's burned flesh had a terrific odor. He was in severe pain but took his experience magnificently. Bill's courage made men weep."

Jansson had seen men weep and had wept himself, but as soon as he saw that the problem was medical and the medical men were there, he was on the move again. He picked two of the rescue crew to accompany him across the ridge and into Mann Gulch to explore ahead for the doctors and be ready to point out where the living and the dead lay hidden. He must have picked out the two he trusted most—one was Don Roos, assistant ranger from the Lincoln District, and the other the seventeen-year-old boy he had met only a few hours before who by now was on his way to prove his own secret belief that he was the best man on the crew.

It was 1:20 A.M. when the three crossed the ridge and started down the other side, where they soon ran into what Jansson describes as a "twelve-foot rim rock breaking off on the Mann Gulch side." Jansson says they had trouble finding a gap in it; others before them had, too.

It would not be exact to say that the three in descending at night into the remnants of Mann Gulch were descending into the valley of the shadow of death, because there was practically nothing left standing to cast a shadow. Since dead trees occasionally exploded and then subsided weakly into dying flames, perhaps it would be more exact to say they were descending into the valley of the candles of death. Rumsey speaks of the night as a "pincushion of fire."

At about 1:50 they heard a cry below and to the right. As they continued to descend, "the updrafts brought a very suspicious smell," but Jansson says that, because the wind was tricky, it was difficult to determine "whether there was a series of bodies ahead or whether we were just smelling Sylvia."

It took them another ten minutes to find Sylvia, probably because Sylvia had been slipping in and out of consciousness during that time.

When Jansson, Roos, and Sallee reached him, Sylvia was standing on a rock slanting heavily downhill. Hunched over and wobbling to keep his balance, he couldn't stop talking. "Please don't come around and look at my face; it's awful." Then he said, "Say, it didn't take you fellows long to get here." He thought it was 5:00 in the morning. Jansson pulled out his watch and said, "It's 2:00 A.M. on the nose." Then in his report, Jansson speaks to us. "Since his hands were burned to charred clubs, I peeled an orange and fed it to him section by section."

Sylvia said, "Say, fellows, I don't think I'll be able to walk out of here." Jansson told him his walking days were over for the time being and he was "going to get a free ride out." He tried to make this a joke, although it is hard to make jokes at night on a hillside that smells of burned flesh.

Sylvia was worried about his shoes, which Dodge had taken off and placed behind a rock, so Jansson combed the slope with a flashlight until he found them. The knowledge that his shoes had been discovered comforted Sylvia, probably because he could not retain knowledge and had slipped back to thinking he would have to walk to the river.

About 2:20 the doctors and most of the rescue crew arrived and treated Sylvia as they had Hellman. Dr. Hawkins agreed with Jansson that it would be dangerous to attempt to move Sylvia and Hellman before daybreak, although the crew was ready to start stumbling in darkness through rocks and reefs to the river.

Sylvia complained of the cold, as Hellman had, but Hellman was wrapped in the only blanket the crew had brought back on its return trip from Hilger Landing. Since most of the men were not wearing jackets, "some of them stripped off their shirts and undershirts to wrap around Joe to keep him warm." As he was still cold, half-naked they huddled close to him.

When he got warm, he got happy again. Several years ago Dr. Hawkins, who treated both Hellman and Sylvia on the ridge and then in the hospital, told me that, if I were burned and wanted to be as happy as Joe Sylvia had been, I should get terribly burned. "Then," he said, "your sensory apparatus dumps into your bloodstream." He added, "Usually it takes until the next day to clog your kidneys. In the meantime, it is possible to have spells when you think you are happy."

Since only two could cuddle close to Sylvia at a time, others of the rescue crew spread out across the hillside looking for eleven missing men by flashlight and candlelight. It was like high mass until dawn—lights walked about all night in darkness.

Sylvia encouraged those who remained with him by telling them that before they had arrived he had heard voices of men calling from above. They were the voices of men working and he had shouted back at them. Perhaps, then, it would be more exact to call Mann Gulch on this night the valley of candles and voices of dead men working.

Daylight came a little after four o'clock, and Jansson walked only a few yards before running into Harrison's body. He identified it by the Catholic medallion around its neck and the snake-bite kit which he had given Harrison when Harrison became recreation guard at Meriwether. His body lay face down pointing uphill and looking as if, instead of being a Catholic, he were a Moslem fallen in prayer. Jansson describes the earth as it looked at daybreak.

> The ground appearance was that a terrific draft of superheated air of tremendous velocity had swept up the hill exploding all inflammable material, causing a wall of flame (which I had observed from below at 5:30 P.M. the previous evening) six hundred feet high to roll over the ridge and down the other side and continue over ridges and down gulches until the fuels were so light that the wall could not maintain heat enough to continue. This wall covered three thousand acres in ten minutes or less. Anything caught in the direct path of the heat blast perished.

Three thousand acres is close to four and three-quarters square miles.

At about 4:40 A.M. they started to carry Sylvia down Mann Gulch to the river. The crew that left with him was only six men and the doctors, so Sallee had to take his turn carrying the litter. It was also up to him to help identify the bodies—they tagged three while carrying Sylvia down the hillside. Jansson, who was noted for being a hard man on himself and his men, was sorry for Sallee. What a great compliment for a seventeen-year-old.

While they continued downhill, Jansson continued to be

puzzled about why Harrison's body had been found so close to Sylvia. He had heard from both Sallee and Rumsey that Harrison had given out from exhaustion, so Jansson had expected to find his body much lower on the hillside and farther back than any of the others. That he got up and climbed to where he did is as much a monument to his courage as the cross they put there afterwards.

Jansson is the only one to have left an account at all inclusive of the discovery, identification, and removal of the bodies. Near each body he left a note under a pile of rocks identifying the body and summarizing the evidence on which the identification had been made. He may have intended to expand these notes into a more complete account, but he never did. If he had tried to say more, it would have been too much, for him and for us.

Lower down the hillside than they thought any of the crew would be found, they came upon Stanley J. Reba's body; but when they examined it, they found he had broken a leg and then no doubt had rolled down the slope into the fire. He had literally burned to death. Most of the others, in all likelihood, had died of suffocation and were burned afterwards.

Sylvia was carried to the mouth of Mann Gulch by Jansson and his crew of six, arriving there only a short time before Hellman reached the river by way of Rescue Gulch, carried by Rumsey and other members of the rescue crew. Neither Sylvia nor Hellman was suffering, because, as Dr. Hawkins adds, "their burns were so deep and hard their nerve ends were destroyed."

Each man was soon picked up by a speedboat, and each man's spirits rose. Sylvia arrived at the hospital in Helena about 10:00 A.M. and Hellman about half an hour later. Dr. Hawkins told me that 10:00 was about time for the kidneys to fail. He immediately ordered an examination, and the report was as expected, "no urine found." There soon came an end to euphoria; both Sylvia and Hellman were dead by noon.

By 1:00, Jansson, who had been in charge of moving Sylvia to the hospital in Helena, was back in Mann Gulch to renew the search with a fresh crew, including Dodge, and a helicopter to fly the bodies to Helena. According to his plans, he should have been there at least three hours earlier, but the "eggbeater," which had been ordered from Missoula, picked him up at 12:30 instead of 9:00. It's hard for the woods and machines to run on the same schedule, and almost never is it the woods that are late.

Jansson had been the first one to be taken in on the helicopter shuttle, and he immediately started uphill tagging bodies. He started where they had found the three at daybreak and then, as he says, worked up the ridge "by contours." He says that he did not have much time to gather up the personal effects scattered around the bodies: "The terrific blast of heat burned all clothing off, releasing non-inflammable effects, which, if not pinned down by the body, were carried as high as one hundred feet farther up the hill." He found watches or the remains of wallets only by rolling a body over.

Late in the afternoon he looked downhill and saw a "charred stump of a man." He already had found the ninth body, "so I didn't count him and didn't go close enough to determine if it was really a remains." He was through for the day, a long day that had begun early the day before. Not until the next morning, the morning of the seventh, were all the remains found.

Only when all the Smokejumpers in his crew had been accounted for did Dodge fly back to Missoula. It is not hard to visualize him, eyes bloody and clothes dirty, as Sallee found him near the top of the ridge after the fire had passed over him, but it takes a moment of thinking to see him as his wife saw him when he stepped down from the plane in Missoula, fastidious as ever except for the tobacco stains at the corners of his mouth. He had five more years to construct a life out of the ashes of this fire.

Jansson had longer to live than Dodge, but those who knew him say he also had great problems rescuing himself. Asked by the Board of Review at what point he had given up being in charge of the rescue, he replied he just couldn't remember. He couldn't remember because he never gave up the charge. For instance, the year of the fire he twice returned to Mann Gulch to check his original observations of the blowup. Afterwards he wrote "Jansson's Ground Check Statement." Having twice walked and rerun his route with a stopwatch in hand, he concluded that his present report "is within two minutes of the time I have shown in previous statements."

In the end, he had to rescue himself from Mann Gulch by asking to be transferred to another ranger district. It had got so that he could not sleep at night, remembering the smell of it, and his dog would no longer come in but cried all night outside, knowing that something had gone wrong with him.

7

Perhaps Jansson's greatest rescue in Mann Gulch occurred later in the year of the fire. Harry Gisborne, the man above all others who made the study of fire a science, was determined to examine Mann Gulch firsthand before winter came and destroyed crucial evidence. His fear of winter was probably accompanied by a fear that he had not long to live, and he had some theories about fire whirls he wanted to test against facts. In particular, he wanted to test a theory he had formed about the cause of the Mann Gulch blowup. So despite a severe heart ailment he was determined to make the trip, and, without letting his close friends or doctor know, he persuaded Jansson, his disciple, to accompany him. Almost literally he was to die for his theory about the cause of the Mann Gulch fire.

The intensity of Gisborne's interest in the cause of the blowup at Mann Gulch and in blowups in general is still another sign of his being an advance-guard scientist. Even as late as the Mann Gulch fire there was no general agreement about the causes of these explosions of wildfires. A blowup is a phenomenon that occurs rarely and often as unpredictably as it occurred that afternoon in Mann Gulch; to add to its secretiveness, it takes place far from the known habitat of meteorologists and trained weather observers. Throughout history, blowups have been seen almost entirely by survivors of big forest fires, who would not have survived if they had stopped to observe them.

Even though Jansson's testimony before the Board had described the Mann Gulch fire as "a blowup," the official *Report of Board of Review* never uses the noun "blowup" or any such adjective as "explosive." Discussion of the behavior of the fire is limited primarily to its appearance as a routine fire prior to the crew's being dropped, perhaps because the Forest Service wanted to downplay the explosive nature of the Mann Gulch fire to protect itself against public charges that its ignorance of fire behavior was responsible for the tragedy. It was not until the 1950s, however, that Clive M. Countryman and Howard E. Graham published articles analyzing fire whirls in wildfires that received general acceptance. And it was only after Laird Robinson and I had taken several trips into Mann Gulch in the late 1970s that we saw clearly how these theories explained the explosive complexity of the Mann Gulch fire.

At the time of the fire or soon after, several of the leading Forest Service scientists stationed in Missoula, such as Jack S. Barrows and Charles E. Hardy, advanced a very different theory as to the cause of the blowup, a theory which still has some standing and from its nature would be very difficult to disprove. This theory is based on the assumption appearing most often when the human mind seeks to explain extraordinary effects—that extraordinary effects must be produced by extraordinary causes. According to this then-prevailing theory, the particular extraordinary cause of the blowup of the Mann Gulch fire was a thunderhead.

Stated simply, this theory presupposes that a thunderhead came along and sat down on the fire—its cool air being heavier than the light, hot air rising from the ground—but the thunderhead never got all the way to the ground in the form of rain. In effect, its sudden weight as it sat down on the top of the fire splattered the fire all around in the form of spot fires, and the gusts of wind that hurry along with dry thunder helped to fan the spot fires and the main fire until in a few minutes it was all fire.

The strongest argument in favor of this theory is that there were highly variable air conditions on this day, which was setting a record for high temperature, and highly variable air conditions and explosions of fire come out of the same bag. The plane ride from Missoula to Mann Gulch had been rough enough to make one Smokejumper get sick and turn in his jumping suit for good. Even Rumsey and Sallee were beginning to feel ill and wanted to be among the first to jump. Furthermore, the pilot reported cumulus cloud formations in the distance at the time the plane was circling the fire, and each of those cumulus puffs signified a heavy updraft of hot air. When columns of hot air reach around twenty-five thousand feet and encounter rain and ice crystals, they are cooled and change to thunderheads. Being now heavier than the hot air around them, they start down but can stop without raining—not, however, without causing a scurry or blast of big winds.

The main trouble with this theory is that none of the survivors mentions a thunderstorm passing by, nor does Jansson, who was in the vortex of the blowup and a casualty of it. Moreover, it is the kind of all-purpose theory you can't disprove that somebody offers whenever a fire blows.

Of course, there is no way in this cockeyed world of ruling out the extraordinary-cause-for-the-extraordinary-effect. You come by boat to Mann Gulch by way of the cliffs of the Missouri River where extraordinary ocean beds stood up and fought each other, but it seems as if the more that becomes known about big cockeyed things, including the actions of men and women as well as cliffs, the more they seem to reduce to one little cockeyed thing fitting closely to another of the same kind, and so on until it all adds up to one big cockeyed thing. It's never confusion, though, because ultimately it all fits—it's just cockeyed and fits and is fire. And of course that is extraordinary.

The extraordinary monster needing explanation is at least in its prenatal form a simple little mechanism. A blowup is a dust kitten that has become a raging monster, but its basic mechanism is that of a swirl of dust that seemingly comes from nowhere and may pick up a loose newspaper and give it a toss. When we think of it as a monster, though, it is natural to think that something out of the sky had to start it spinning, and it is probable that some blowup somewhere was started by a thunderhead making a big wind spinning in circles, and it is proper, in searching for the cause of blowups, to consider the thunderhead theory. But the other basic theory of the origin of blowups, and the one we shall be dealing with, can be called the "obstacle theory." It is the theory of Countryman and Graham and the theory that has met general acceptance. And, not surprisingly, it is Gisborne's underlying theory, although he had not developed it sufficiently to explain the Mann Gulch fire correctly. It was not until Laird and I returned to Mann Gulch on hot mornings and continued to puzzle over what we saw there that we began to notice each time the same combination of little things that would fit together to start a fire whirl if, as was the actual case, a fire were present near the mouth of the gulch on its southern side and near the top of the ridge. Among woodsmen there is a preference for causes that are there waiting for you when you return, but admittedly sometimes they drop from the sky.

The obstacle theory in its essential elements is not hard to understand. A wind strikes an obstacle, say a rocky promontory on a ridge, shears off it, and so starts to spin and soon goes into full circles behind the promontory. Any fire caught in these circles will throw off sparks and even burning branches which, if the conditions are right, will start spot fires, and these, when the conditions continue to be favorable, will swell into fire swirls, and when you get caught between them and the main fire you will be as lucky as Jansson if you regain consciousness in time to vomit. All this is easy to visualize if you like to walk by moving waters and note what happens in a stream when it strikes a half-submerged rock or small logjam. The stream shears off it, and the good fishing is where the eddies form on the rear flanks of the obstacle and behind it. No trouble at all for stream fishermen to visualize.

Soon the question of how a strike of lightning in a dead snag high up near the top of a ridge close to the mouth of Mann Gulch became a fire monster consuming Mann Gulch and thirteen elite firefighters turns into the question, Where are the winds of yesteryear? And that poetic question soon turns up the equally poetic answer, Gone with the winds. And that poetic question and answer when translated into direct prose means that you can't explain the cause of a big fire of long ago if you can't reconstruct the winds that caused it, and also that nothing is more true than that each individual wind passes and is gone for good. But the practical woodsman, who seldom is a poet, starts with the assumption that at least some of the winds of yesteryear are not gone, if only one knows how to see a wind that is gone. The practical woodsman thinks that he can see a lot of things in the woods that will tell him a lot about what can't be seen there. For instance, you may already have guessed how much Laird and I explain what we see in the woods by relying on what we have seen when fishing. It shouldn't be surprising, then, that an important part of our theory of what caused the Mann Gulch blowup was an observation we made from a boat on the Missouri River several miles before getting to the mouth of Mann Gulch.

We were on our first trip together to Mann Gulch, in 1977, and I had been left behind not happily at the mouth of the gulch, where there isn't much to do or see on a hot August afternoon. Laird clearly and if anything overpolitely had left me behind. He had a theory to check that would take him sidehilling to the head of Mann Gulch, and the unspoken word was that if I went along I would slow him down. He left on a supposedly cheery note to the effect that, while he was killing himself on the hot, bare hillside, I could loaf around the mouth of the gulch, which the river left cool, with plenty of time to find a missing part of the puzzle of what caused the blowup. And, so help me, I more or less did.

Nearly thirty years after a fire has burned over a piece of shale in the Gates of the Mountains, there doesn't seem to be much to see since almost no trees are left standing; black fallen trees thirty years after don't seem to offer many opportunities to make contributions to knowledge. I thought to myself, "Maybe you are trying to see something big and important too soon. Maybe it would be surer to come if you tried to work up to it." So I backed off, with only one slightly odd thing to observe about the mouth of the gulch—that there are a few green and standing trees there, just a short stretch of them, a hundred yards or so of them between the rise and the edge of the old fire, and I thought to myself, "That must have been a hell of a big wind to blow all this fire upgulch after it jumped the canyon. You would have thought a little of the fire would have sneaked a short way backward and toward the river."

I started walking up the canyon, slowly, very slowly. Above the mouth of the gulch there seemed to be nothing to observe but black fallen trees, and after thirty years of lying on the ground they look pretty much alike. After I reconciled myself to the fact that all I was going to see was black fallen trees, I finally said to myself, "About all that is left for me to see is the way the dead trees fell," and to my astonishment I just then saw something—or at least something that might be something.

Remember, now, that when I was looking at the way trees had fallen I was really looking for winds that had gone, and almost immediately I saw that the black, dead bodies of the fallen trees on the southern side of the gulch where the fire had started were strangely parallel to each other but at right angles to the top of the ridge. My immediate reaction was Everyman's reaction. I turned and looked to the top of the ridge on the opposite, or northern, side of the gulch where there were also dead fallen trees—lots of them—and in a pattern, too, but in a puzzling one. They were lying parallel to each other, but, unlike the trees on the southern side, they lay parallel to the top of the ridge. However puzzling the patterns, the patterns had to stand as the remains of winds.

They probably had to stand for prevailing winds and for winds that might still be there. Certainly they had to have been there for some years after the trees burned, long enough for the trees to have rotted in the roots and been blown over. One vast storm might have done it but not likely—the trees couldn't have rotted uniformly and agreed to topple at the same time. They had to have been worked on fairly regularly over the years. As prevailing winds, they might still be there at their more or less appointed time, although the pattern of winds might since have changed. But it was fairly sure that once and for some years a big wind had blown over the top of the southern ridge and then down it (at right angles to the top of the ridge) and that on the northern side a big wind with some regularity had blown parallel to the ridge near its top.

This was the best I could do until Laird got back from his mission to the head of the gulch, but at first he wasn't much interested in my report. His own report left him fairly depleted. He told me that we needed a new theory to explain why most of the crew, after leaving the escape fire, kept side-hilling up the gulch instead of going for the ridge. We had imagined a long stretch of impregnable reef blocking their escape. "In fact," Laird said, "there were several big openings in the reef that they passed by but could easily have crossed through."

We both felt depleted by this negative report. To spend a day in Mann Gulch, we had had to drag a motorboat on a trailer 130 miles over the Continental Divide just to get to the Missouri River. From there to where we were now in the late afternoon had taken the rest of the day, and it was about time to fold up and start back up the river to get to Missoula not too long after midnight. About all we would be able to show for a long day in Mann Gulch was that one of our theories about the tragedy was proven wrong by the hardest of evidence—the ground. So my report about the mess of burned, fallen trees on opposite sides of the gulch wasn't going to take away our disappointment over the results of our long day. But it was about all we had to show for it, and I made my report brief. Still, neither of us entirely forgot it. We talked about it at several of our customary lunches in Missoula, and it was not long before the parallel messes began to emerge as something that might be important.

It was only a trip or two later that we started to think of the Missouri River as having a possible connection with the blowup of the Mann Gulch fire. Up to this time, the Missouri River had been scenically interesting to us, but mostly it had meant motor trouble for our boat. We usually spent as much time on the river trying to figure out what the missing parts of our motor were as we did trying to figure out the missing parts of the story of the fire. On this day we had gone nearly a mile before the motor stopped, so we were still roughly five miles from the mouth of Mann Gulch. While Laird was kicking the motor, trying to get it to start a second time, I was trying to size up this piece of the river we were floating on to figure out how I would fish it. But no matter how much I was thinking about something else, even fishing, I was always ready to think about prevailing winds, especially when I got anywhere near Mann Gulch. I had noted a medium-sized wave in the quiet water near shore, and I thought to myself, "That's funny."

All I meant by "funny" is that the wave was going the wrong way from the way I thought it should be going or, more exactly, the wind blowing it was going the wrong way for a prevailing wind on a big mountain river at this time of day. This early in the day in hot summer, the prevailing wind on a mountain river should be blowing upstream as follows. The rising sun hits the tops of mountains first. The warming mountain air, being lighter than the cool valley air, will rise, and the valley air will rush upstream to take its place. In the late afternoons or evenings, it is generally the other way around—the tops of mountains cool first, cool air is drawn into the warmer valley below, and the prevailing wind is usually downstream. From here on I kept watching the waves on the river and trying to connect them with Mann Gulch.

I wasn't able to make that connection immediately, but I was able to find a reasonable explanation for the odd downstream wind that blows in the Gates of the Mountains on hot mornings. The air deep in the Gates of the Mountains is always much cooler than the air on the plains outside the Gates. As the river passes through the cooler air inside the cliffs to the plains, the hot plains air rises and the cool air is drawn downstream to replace it. Accordingly, in the cliffs of the Gates of the Mountains, the prevailing wind in the morning and early afternoon is downstream, just as it is in the late afternoons and evenings when the mountains cool.

Coming downriver and thinking about prevailing winds as I was, you see ahead several great bends in the river so complete they look from a distance as if the river ahead has run into a mountain and disappeared under it. Captain Meriwether Lewis, coming upriver, must several times have wondered if ahead he wasn't about to run out of river, leaving him and his crew with boats and oars but nowhere to row to. It must have been an overwhelming sight to come suddenly upon what seems to be the abrupt end of a big river several hundred miles before it should end.

The explanation for this mirage only slowly emerges as the mountain ahead under which the river disappears takes shape. Slowly, it emerges as a promontory sticking out into the river and causing the river to make a sharp bend to the northwest. It is an obstacle that turns the river at the mouth of Meriwether Canyon, just upriver from Mann Gulch. The promontory begins at the mouth of the canyon where Lewis and his crew camped overnight and extends half a mile before allowing the river to pass around it. As the river bends north again, the waters swirl with considerable force into the mouth of Mann Gulch just below the promontory. After that, we don't care anymore where the river goes—to the Mississippi supposedly.

For several miles we have been following a downriver wind wondering if it could by chance have any connection with Mann Gulch. The mountain ahead becomes more interesting when it turns into a promontory between Meriwether Canyon and Mann Gulch than when it was a mirage under which a river disappeared. As a promontory, it supplies the connection between the prevailing downriver wind on the Missouri and the blowup of the Mann Gulch fire.

When the prevailing downriver wind (call it Wind Num-

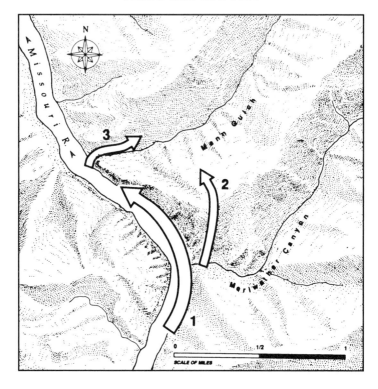

Map of wind directions in Mann Gulch on August 5, 1949.

ber One) first struck the promontory near the mouth of Meriwether Canyon, it split as if it had struck a big rock in a river and part of it (Wind Number Two) went straight ahead over the promontory and then straight down it in eddies on the Mann Gulch side. Some eddies became fire whirls throwing spot fires, and soon some of these spot fires jumped Mann Gulch to the opposite, grassy, slope. It was no great jump, the gulch being narrow and its grasses on the north side cured and already heated by radiation. Wind Number One, the main downriver wind, continued around the promontory until it met the mouth of Mann Gulch and sent Wind Number Three surging upgulch. Wind Number Three

struck Wind Number Two at right angles, creating a giant
fire whirl and starting the fateful race up Mann Gulch. As
the fire swirled upgulch, a convection effect added strength
to Wind Number Three: cool air from the river rushed up
the narrow funnel at the mouth of Mann Gulch to replace
the hot air rising from the fire.

By now it should be clear that an important part of the
story of the Mann Gulch blowup is a Story of Three Winds.
These three winds not only help to explain the basic structure of the blowup; they also explain other details. For instance, the powerful upgulch wind (Wind Number Three)
struck the burned fallen trees on the north side of the gulch;
they lie parallel to each other and also parallel to the top
of the ridge. The burned fallen trees opposite, on the south
side of the gulch, are the work of Wind Number Two. This
is the wind that crossed the promontory and headed almost
straight down to the bottom of Mann Gulch. It would explain why the burned fallen trees are parallel to each other
but perpendicular to the top of the ridge. There is probably
an upgulch wind in Mann Gulch much of the summertime.
What we have called Wind Number Three only added power
to it, but power it needed to have. It was the wind that ran
its race against the Smokejumpers and won.

It must be remembered, though, that these three winds
are all part of one, first observed as a wave on the Missouri
going the wrong way. Although Wind Number One did not
act directly on the fire, it furnished the power for the other
two that did. It was, so to speak, the supply ship, the other
two the destroyers. As a Story of Three Winds, it solves still
other puzzles the fire left behind. When Jansson was with
Gisborne on the day of Gisborne's death in Mann Gulch, Gisborne saw that the track of the blowup headed straight toward
the part of the ridgetop where Hellman had been burned.

Occasionally in life, there come times that mark the end of puzzles. As was said earlier in this chapter, it is all cock-eyed and it all fits.

———

Although blowups were not analyzed to the satisfaction of fire behavior scientists until the 1950s, it would be almost an *a priori* certainty that Harry Gisborne had long been interested in them. He had both a general theory about what caused them and a corollary theory as to what specifically caused the Mann Gulch blowup. He was to discover in Mann Gulch on the last day of his life that both his theories were wrong. To his credit, though, he was the first one to point out his error and was happily preparing to wake up the next day to correct his theories and probably soon would have, since his theories were basically on the right track.

He was a made-to-order picture of an early scientist, somewhat poetic and sometimes wrong, but often right and nearly always on the right track and nearly always dramatic. He tended to look for the extraordinary cause to explain the extraordinary effect, and, being himself an extraordinary effect, he had gathered about him a cult of young rangers, Jansson among them, to check in the field the theories he had developed at the Priest River Experiment Station, of which he had been director since 1922 and which he had made into the center of Forest Service research in the Northwest.

It is an extraordinary thing to say about a great woodsman, but he must also have been a charming man. He even developed the charming theory that certain conditions could be observed in November from which accurate predictions could be made about the fire season in the coming summer. He almost lost his cult of hero-worshiping rangers when he tried to get them to spend their Novembers looking for signs of next summer's fires. They told him they still loved him even though he must be going crazy.

But they stuck with him to check his theory that fire whirls always whirl clockwise, a theory that also turned out to be wrong—some whirl counter- and some clockwise. But the underlying assumption of the theory is right. Most fire whirls are caused by winds shearing off obstacles, that is, glancing off a side of them and so giving fires a spin and starting them to whirl.

His corollary theory about the cause of the Mann Gulch fire whirl and blowup makes this basic assumption. In a memorandum dated the August 30 following the fire, he urges his friend C. S. Crocker, chief of the Division of Fire Control, to instruct his fire dispatchers and spotters hereafter not to jump men on a fire if it "is so located that air will be sucked into it around a promontory to the left of the fire." When you see any such fire, he says, "look out for a blowup caused by a big clockwise whirl."

The two big blowups he had observed previously worked this way, he affirms, and so had all blowups observed by others he had questioned. And this is no doubt true; he just had too few cases on which to build a general theory.

For his theory to explain the blowup in Mann Gulch, there had to be a promontory in front of (upgulch from) the advancing fire and to its left as you face it, and in addition there had to be a wind blowing downgulch. The downgulch wind would shear off the promontory and strike the main fire behind it on its left side, starting a fire whirl with a clockwise motion, which would throw burning cones and branches outside the circumference of the main fire, where some of them would start spot fires in unburned fuel.

This is fancy and intriguing, like a lot of early science, but Gisborne was a fine enough scientist to know after a good

look at Mann Gulch that it wouldn't work there, for two very
good reasons. First, the only promontory was behind the advancing fire, not in front of it. Second, all the survivors speak
of an upgulch wind, not a downgulch wind, the other necessary ingredient of Gisborne's theory. In Mann Gulch the fire
whirl had begun in a way Gisborne had never seen—with
an upgulch wind shearing off an obstacle behind the fire.
If anything, Gisborne seemed exhilarated by the prospect of
having to get another theory to replace the old one.

It was November 9 when they climbed up Rescue Gulch,
Gisborne half-convincing Jansson that if he rested every
hundred yards he could make it. Jansson says that the half-
hour trip took two hours, and he leaves a strange and moving report (for insurance purposes) about the whole day entitled "Statement to Accompany Form CA2 in the Case of the
Death of Harry T. Gisborne." The report was accompanied by
a map (no longer to be found) with thirty-seven numbers on
it signifying points where the two had stopped to talk. Still
surviving, however, are four pages of notes indicating the
main topics of conversation at each "rest stop" and often a
direct quotation from the conversation there. They had not
gone halfway through the fire area—only to stop 15—when
the notation reads: "All his theories on the fire blasted."

At stop 28, Gisborne said to Jansson: "I don't believe your
fire whirled. You only thought so because of what you have
been told, by me in part."

Gisborne must have been a fairly complex traveling companion. Much as he had been compelled to Mann Gulch by
his theory about the cause of the fire whirl in the Mann Gulch
fire, here he was, before the afternoon was over, half kidding
and half scolding Jansson for believing that there had been
a blowup in Mann Gulch, even though Jansson had almost
died in it. He was even kidding Jansson for being led into

this supposed error by being too subservient to Gisborne's own theories.

To Gisborne, science started and ended in observation, and theory should always be endangered by it. Toward the end of the afternoon all he had observed showed that the fire had burned without fire whirls and that therefore both he and Jansson were wrong. Jansson was barely hanging on to his own experience of the fire whirl when they reached stop 32. Suddenly Jansson said, "That's my whirl." Gisborne immediately saw that a fire whirl had gone on a line for the top of the ridge near where Hellman had been burned, and probably was the reason Hellman had been burned there and Diettert was burned a little farther on. "Yep," he said, and was all fire whirl himself again. He wanted to take pictures of it and then and there in late afternoon to follow its course and map it. Jansson was afraid that they already had made too long a day of it, and he knew that it would be all they could do to get back to their truck before dark. He assured Gisborne he would have someone return to the gulch to map the whirl, and Gisborne, when he saw how perturbed Jansson was, apologized. He said to Jansson: "I'm glad I got a chance to get up here. Tomorrow we can get all our dope together and work on Hypothesis Number One. Maybe it will lead to a theory." This was at rest stop 35. By now the rest stops were becoming stations of the cross.

They were following a game trail along the cliffs high above the Missouri River at the lower end of the Gates of the Mountains, and were only a quarter or a half mile from their truck when they reached stop 37. Gisborne sat down on a rock and said: "Here's a nice place to sit and watch the river. I made it good. My legs might ache a little, though, tomorrow."

In his report Jansson says: "I think Gisborne's rising at point 37 on the map was due to the attack hitting him." He goes on to explain in parentheses that "thrombosis cases usually want to stand or sit up because of difficulty in breathing." Gisborne died within a minute, and Jansson piled rocks around him so he would not roll off the game trail into the Missouri River a hundred feet below.

When Jansson knew Gisborne was dead, he stretched him out straight on the game trail, built the rocks around him higher, closed his eyes, and then put his glasses back on him so, just in case he woke up, he could see where he was.

Then Jansson ran for help. The stars came out. Nothing moved on the game trail. The great Missouri passing below repeated the same succession of chords it probably will play for a million years to come. The only other motion was the moon floating across the lenses of Gisborne's glasses, which at last were unobservant.

This is the death of a scientist, a scientist who did much to establish a science. On the day of his death he had the pleasure of discovering that his theory about the Mann Gulch blowup was wrong. It would be revealing if tomorrow had come and he had got all his dope together and had worked out a new Hypothesis Number One. Maybe it would have led to another theory, probably the right one.

In any case, because of him we have been able to form what is likely the correct theory. Gisborne's portrait hangs on the staircase of the Northern Forest Fire Laboratory in Missoula, which immediately adjoins the Smokejumper base. He looks you square in the eye but is half amused as if he had caught you too attached to one of your theories, or one of his.

This was also the end of Jansson's rescue work in Mann Gulch. He was to be transferred from the Canyon Ferry Ranger Station because he could not sleep in it or get his dog to come inside.

For a scientist, this is a good way to live and die, maybe

the ideal way for any of us—excitedly finding we were wrong and excitedly waiting for tomorrow to come so we can start over, get our new dope together, and find a Hypothesis Number One all over again. And being basically on the right track when we were wrong.

Later, in thinking I was following him, I came to find out much of what I had found missing in Mann Gulch.

PART TWO

8

We enter now a different time zone, even a different world of time. Suddenly comes the world of slow-time that accompanies grief and moral bewilderment trying to understand the extinction of those whose love and everlasting presence were never questioned. All there was to time were the fifty-six speeding minutes before the fire picked watches off dead bodies, blew them up a hillside ahead of the bodies, and froze the watch hands together. Ahead now is a world of no explosions, no blowups, and, without a storyteller, not many explanations. Immediately ahead we know there is bound to be a flare-up of public indignation and the wavering candlelight of private grief. What then? It could be a slow fade-out of slow-time until all that's left of the memory of Mann Gulch are cracking concrete crosses on an almost inaccessible hill and a memorial tablet with the names to go with the crosses beside a picnic ground at the mouth of the next gulch upriver.

After the autumn rains changed ashes into mud slides, the story seems to have been buried in incompleteness, pieces of it altogether missing. As a mystery story, it left unexplained what dramatic and devastating forces coincided to make the best of young men into bodies, how the bodies got to their crosses and what it was like on the way, and why this catastrophe has been allowed to pass without a search for the carefully measured grains of consolation needed to transform catastrophe into tragedy. It would be natural here, looking for at least chronological continuity to the story, to follow the outcries of the public and at the same time to try to share some part of the private sufferings of those who loved those who died. But always it would have to be conceived as possible, if an ending were sought in this direction, that there might be a non-ending. It is even conceivable that most of those closely connected with the catastrophe soon tried to see that it got lost; when the coming controversies and legal proceedings were added to all the rest of it, clearly the whole thing got so big it frightened people. They wanted it to go away and not come back.

Even so, there may somewhere be an ending to this story, although it might take a storyteller's faith to proceed on a quest to find it and on the way to retain the belief that it might both be true and fit together dramatically. A story that honors the dead realistically partly atones for their sufferings, and so instead of leaving us in moral bewilderment, adds dimensions to our acuteness in watching the universe's four elements at work—sky, earth, fire, and young men.

True, though, it must be. Far back in the impulses to find this story is a storyteller's belief that at times life takes on the shape of art and that the remembered remnants of these moments are largely what we come to mean by life. The short semi-humorous comedies we live, our long certain tragedies, and our springtime lyrics and limericks make up most of what we are. They become almost all of what we remember of ourselves. Although it would be too fancy to take these moments of our lives that seemingly have shape and design as proof we are inhabited by an impulse to art, yet deep within us is a counterimpulse to the id or whatever name is presently attached to the disorderly, the violent, the catastrophic both in and outside us. As a feeling, this counterimpulse to the id is a kind of craving for sanity, for things belonging to each other, and results in a comfortable feeling when the universe is seen to take a garment from the rack that seems to fit. Of course, both impulses need to be present to explain our lives and our art, and probably go a long way to explain why tragedy, inflamed with the disorderly, is generally regarded as the most composed art form.

It should be clear now after nearly forty years that the truculent universe prefers to retain the Mann Gulch fire as one of its secrets—left to itself, it fades away, an unsolved, violent incident grieved over by the fewer and fewer still living who are old enough to grieve over fatalities of 1949. If there is a story in Mann Gulch, it will take something of a storyteller at this date to find it, and it is not easy to imagine what impulses would lead him to search for it. He probably should be an old storyteller, at least old enough to know that the problem of identity is always a problem, not just a problem of youth, and even old enough to know that the nearest anyone can come to finding himself at any given age is to find a story that somehow tells him about himself.

When I was a young teacher and still thought of myself as a billiards player, I had the pleasure of watching Albert Abraham Michelson play billiards nearly every noon. He was by then one of our national idols, having been the first American to win the Nobel Prize in science (for the measurement of the speed of light, among other things). To me, he took on added luster because he was the best amateur billiards player I had ever seen. One noon, while he was still shaking his head at himself for missing an easy shot after he had had a run of thirty-five or thirty-six, I said to him, "You are a fine billiards player, Mr. Michelson." He shook his head at himself and said, "No. I'm getting old. I can still make the long three-cushion shots, but I'm losing the soft touch on the short ones." He chalked up, but instead of taking the next shot, he finished what he had to say. "Billiards, though, is a good game, but billiards is not as good a game as chess." Still chalking his cue, he said, "Chess, though, is not as good a game as painting." He made it final by saying, "But painting is not as good a game as physics." Then he hung up his cue and went home to spend the afternoon painting under the large tree on his front lawn.

It is in the world of slow-time that truth and art are found as one.

———

The hill on which they died is a lot like Custer Hill. In the dry grass on both hills are white scattered markers where the bodies were found, a special cluster of them just short of the top, where red terror closed in from behind and above and from the sides. The bodies were of those who were young and thought to be invincible by others and themselves. They were the fastest the nation had in getting to where there was danger, they got there by moving in the magic realm between heaven and earth, and when they got there they almost made a game of it. None were surer they couldn't lose than the Seventh Cavalry and the Smokejumpers.

The difference between thirteen crosses and 245 or 246 markers (they are hard to count) made it a small Custer Hill, with some advantages. It had helicopters, air patrol, and lawsuits. It had instant newspaper coverage and so could heighten the headlines and suspense as the injury list changed from three uninjured survivors, two men badly burned, and the rest of the crew missing, to the final list on which the only survivors were the first three. The headlines flamed higher as all the burned and all the missing proved dead.

The Forest Service knew right away it was in for big trouble. By August 7 chief forester Lyle F. Watts in Washington appointed an initial committee to investigate the tragedy and to report to him immediately, and on August 9 the committee flew over the area several times, went back to Mann

Gulch by boat, and spent three hours there. Like General Custer himself, who liked to have reporters along, the investigating committee took with them a crew of reporters and photographers from *Life* magazine. The lead article of *Life*'s August 22, 1949, issue, "Smokejumpers Suffer Ordeal by Fire," runs to five pages and includes a map and photographs of the fire, the funeral, and a deer burned to death, probably the deer that Rumsey and Sallee saw come out of the flames and collapse while they were ducking from one side of their rock slide to the other.

Accompanying the Smokejumpers on their August 5 flight was a Forest Service photographer named Elmer Bloom, who had been commissioned to make a training film for young jumpers; Bloom took shots of the crew suiting up and loading, of the Mann Gulch fire as it was first seen from the plane, and of what was to be the last jump for most of the crew. There are five frames from his documentary reproduced in *Life*, and for all my efforts to find the film, this is all I have ever seen of it. I did find an August 1949 letter from the regional forester in Missoula to the chief forester in Washington saying in effect that the film was too hot to handle at home so he was sending it along. Nobody in Washington can find it for me. I'm always told it must have been "misfiled," and it may well have been, since there is no better way in this world to lose something forever than to misfile it in a big library.

It is hard to believe the film could be anything other than an odd little memento, but very early the threats of lawsuits from parents sounded in the distance and the Forest Service reversed its early policy of accommodating *Life*'s photographers to one of burying the photographs they already had.

The leader of the public outcry against the Forest Service was Henry Thol, the grief-unbalanced father of Henry, Jr., whose cross is closest to the top of the Mann Gulch ridge. Not only was the father's grief almost beyond restraint, but he, more than any other relative of the dead, should have known what he was talking about. He was a retired Forest Service ranger of the old school, and soon after the fire he was in Mann Gulch studying and pacing the tragedy. Harvey Jenson, the man in charge of the excursion boat taking tourists downriver from Hilger Landing to the mouth of Mann Gulch, became concerned about Thol's conduct and the effect it was having on his tourist trade. His mildest statement: "Thol has been very unreasonable about his remarks and has expressed his ideas very forcefully to boatloads of people going and coming from Mann Gulch as he rides back and forth on the passenger boat."

The Forest Service moved quickly, probably too quickly, to make its official report and get its story of the fire to the public. It appointed a formal Board of Review, all from the Forest Service and none ranked below assistant regional forester, who assembled in Missoula on September 26, the next morning flew several loops around and across Mann Gulch, spent that afternoon going over the ground on foot, and during the next two days heard "all key witnesses" to the fire. The *Report of Board of Review* is dated September 29, 1949, three days after the committee members arrived in Missoula, and it is hard to see how in such a short time and so close to the event and in the intense heat of the public atmosphere a convincing analysis could be made of a small Custer Hill. In four days they assembled all the relevant facts, reviewed them, passed judgment on them, and wrote what they hoped was a closed book on the biggest tragedy the Smokejumpers had ever had.

In this narrative the *Report* and the testimony on which it was based have been referred to or quoted from a good many times, and an ending to this story has to involve an examination of the *Report*'s chief findings. But the immediate effect of the Forest Service's official story of the fire was to add to its fuel.

By October 14, Michael ("Mike") Mansfield, then a member of the House of Representatives and later to be the distinguished leader of the Democratic senators, pushed through Congress an amendment to the Federal Employees' Compensation Act raising the burial allowance from two hundred dollars to four hundred dollars and making the amendment retroactive so it would apply to the dead of Mann Gulch.

The extra two hundred dollars per body for burial expenses did little to diminish the anger or personal grief. By 1951, eight damage suits had been brought by parents of those who had died in the fire, or by representatives of their estates, "alleging negligence on the part of Forest Service officials and praying damages on account of the 'loss of the comfort, society, and companionship' of a son." To keep things in their proper size, however, it should be added that the eight suits were filed by representatives of only four of the dead, each plaintiff bringing two suits, one on the plaintiff's own behalf for loss of the son's companionship and support, and the other on behalf of the dead son for the suffering the son had endured.

Henry Thol was the leader of this group and was to carry his own case to the court of appeals, so a good way to get a wide-angle view of the arguments and evidence these cases were to be built on is to examine Thol's testimony before the Board of Review, at which he was the concluding witness.

Thol had become the main figure behind the lawsuits not just because he alone of all the parents of the dead was a lifelong woodsman and could meet the Forest Service on its own grounds. Geography increased his intensity: he lived in

Kalispell, not far from the Missoula headquarters of Region One and the Smokejumper base. In addition, Kalispell was the town which had been Hellman's home. Thol knew Hellman's wife, and his grief for her in her pregnancy and bewildered financial condition only intensified his rage. There is also probably some truth in what Jansson told Dodge in a sympathetic letter written after the fire: Jansson says that Thol, like many old-time rangers, felt he had been pushed around by the college graduates who had taken over the Forest Service, so he was predisposed to find everything the modern Forest Service did wrong and in a way designed to make him suffer. In his testimony every major order the Forest Service gave the crew of Smokejumpers is denounced as an unpardonable error in woodsmanship—from jumping the crew on a fire in such rough and worthless country and in such abnormal heat and wind, to Dodge's escape fire. The crew should not have been jumped but should have been returned to Missoula; as soon as Dodge saw the fire he should have led the crew straight uphill out of Mann Gulch instead of down it and so dangerously paralleling the explosive fire; Dodge should not have wasted time going back to the cargo area with Harrison to have lunch; when Dodge saw that the fire had jumped the gulch and that the crew might soon be trapped by it, he should have headed straight for the top of the ridge instead of angling toward it; and so on to Dodge's "escape" fire, which was the tragic trap, the trap from which his boy and none of the other twelve could escape.

The charge that weighed heaviest with both Thol and the Forest Service from the outset was the charge that Dodge's fire, instead of being an escape fire, had cut off the crew's escape and was the killer itself.

While the fire was still burning, the Forest Service was alarmed by the possibility that the Smokejumpers were burned by their own foreman. The initial committee of investigation appointed by the chief forester in Washington was headed by the Forest Service's chief of fire control, C. A. Gustafson, who testified later that he had not wanted to talk to any of the survivors before seeing the fire himself because of certain fears about it which he wanted to face alone. In his words, this is what brought him into Mann Gulch on August 9. "The thing I was worried about was the effect of the escape fire on the possible escape of the men themselves."

Jenson, the boatman who took tourists downriver to Mann Gulch, makes probably the most accurate recording of Thol's opinion of Dodge's fire. He said he had heard Thol say, "There is no question about it—Dodge's fire burned the boys." And the father didn't blur his words before the Board of Review. "Indications on the ground show quite plainly that [Dodge's] own fire caught up with some of the boys up there above him. His own fire prevented those below him from going to the top. The poor boys were caught—they had no escape."

The Forest Service's reply to this and other charges can be found in the conclusions of the *Report of Board of Review.* The twelfth and final conclusion of the *Report* is merely a summary of the conclusions that go before it: "It is the overall conclusion of the Board that there is no evidence of disregard by those responsible for the jumper crew of the elements of risk which they are expected to take into account in placing jumper crews on fires."

The Board also added a countercharge: the men would all have been saved if they had "heeded Dodge's efforts to get them to go into the escape fire area" with him. Throughout the questioning there are more than hints that the Board of Review was trying to establish that a kind of insurrection occurred at Dodge's fire led by someone believed to have said, as Dodge hollered at them to lie down in his fire, "To hell with that, I'm getting out of here!" As we have seen, there were even suggestions that the someone believed to have said this was Hellman and that, therefore, the race for the ridge was triggered by the second-in-command in defiance of his commander's orders.

Rumors still circulate among old-time jumpers that there was bad feeling between Dodge, the foreman, and his second-in-command, Hellman, although I have never found any direct evidence to support such rumors. There seems somehow to be a linkage in our minds between the annihilation near the top of a hill of our finest troops and the charge that the second-in-command didn't obey his leader because there had long been "bad blood between them." Custer's supporters explain Custer's disaster by the charge that his second-in-command, Major Reno, hated the general and in the battle failed to support him, and it is easy to understand psychologically how a narrative device like this can become a fixed piece of history. It relieves "our" leader and all "our" men from the responsibility of having caused a national catastrophe, all except one of the favorite scapegoats in history, the "second guy."

Back in the world where things can be determined, if not proved, the suit of Henry Thol reached the United States Court of Appeals for the Ninth Circuit, where Warren E. Burger, later chief justice of the United States Supreme Court, was one of the lawyers arguing against the retired United States Forest Service ranger's charge that his son would have made the remaining sixty or seventy yards to the top of the ridge except for the negligence of the United States Forest Service. Despite the number of times "United States" occurs in that sentence, nothing much happens because of it.

The court of appeals in 1954 upheld the decision of the district court for the district of Montana. Both courts therefore ruled that the Federal Employees' Compensation Act could constitutionally exclude nondependent parents from receiving damages from the federal government beyond the burial allowance. The decision was also upheld that Mansfield's amendment was constitutional in making itself retroactive so that it would apply to the dead of Mann Gulch. Therefore, after the parents had buried their boys, they received another two hundred dollars.

———

For a time it looked as if the four hundred dollars would put an end to the story. A court decision built on two hundred dollars plus two hundred dollars per body silenced the parents; they could not pursue their charge of negligence unless the decision of the court of appeals was reversed by our highest court. To the average citizen the government holds nearly all the cards and will play them when the government is both the alleged guilty party and judge of its own guilt.

For instance, late in 1951 (December 12) in Lewiston, Idaho, Robert Sallee made a second statement about the Mann Gulch fire to "an investigator for the United States Forest Service," and less than a month later (January 1, 1952) Walter Rumsey in Garfield, Kansas, made his second statement to the same official investigator. The Forest Service was preparing its two key witnesses just in case, going so far as to bring them back to Mann Gulch for a refresher course. Their second statements follow closely their first, which were taken only a few days after their return to Missoula from the fire (both dated August 10, 1949). In fact, their later statements follow their first statements word for word a good part of the way, so the alterations stick out like sore thumbs, which is still a good figure of speech at times. Two assurances were drawn from Rumsey and Sallee: they now protested at length that if those who died had followed Dodge's appeals to lie down with him in his own fire (as they themselves hadn't), they, like Dodge, would have been saved. Moreover, they now insisted at length that in escaping over the hill they had followed the upgulch edge of Dodge's fire "straight" to the top of the ridge and so Dodge's fire could not have pursued those who were burned toward the head of the gulch, since they and his fire would have been going roughly at right angles to each other.

Some official documents about the fire, then, were re-touched and given the right shading. More of them probably were just buried—some were even marked "Confidential" and were held from the public as if these fire reports endangered national security. Still others were scattered among different Forest Service offices from the headquarters of Region One in Missoula to the national headquarters in Washington, D.C. The Mann Gulch fire was so scattered that about a quarter of a century later the first mystery I had to solve about it was to find out where it had disappeared. It was like a burial at sea—it was hard afterwards to find the bodies in the wash, and I couldn't have done it without the help of men and women of the Forest Service who felt the burial had been indecent. It's a different world now anyway from what it was before the passage of the Freedom of Information Act in 1966, but I lived most of my life before that date and I can remember it being a loud laugh at Regional Headquarters in Missoula when an energetic, outspoken journalist tried persistently to get access to the documents there on the Mann Gulch fire. With eight lawsuits in the docket brought against the Forest Service for negligence in the deaths of thirteen of the West's finest, the Forest Service was not opening its heart or its files for all to see. As far as the files marked "Mann Gulch Fire" were concerned, mum was the word.

The reasons why parents, relatives, and close friends hoped for silence are naturally very different from the government's. The Forest Service sought silence; the parents were reduced to it, although in sad ways they may also have sought it. On the whole, they were not people of means and could not afford to appeal their case to the Supreme Court, even if they had wished, and, except for Thol, they must have had small understanding of their own case and therefore an underlying reluctance to pursue it. Most important of all probably is the secrecy of the grief and moral bewilderment suffered at the death of one of ourselves who was young, had a special flair, a special daring, a special disregard for death, who seemed, both to himself and to us, to be apart from death, especially from death leaving behind no explanation of itself either as a sequence of events or as a moral occurrence in what-kind-of-a-universe-is-this-anyway. It is the frightened and recessive grief suffered for one whom you hoped neither death nor anything evil would dare touch. Afterwards, you live in fear that something might alter your memory of him and of all other things. I should know.

A few summers ago, thirty years after the fire, I sent what I hoped was gentle word through a common friend to a mother of one of the Mann Gulch dead asking if I could talk to her, and she returned through the common friend gentle word saying that even after all these years she was unable to talk about her son's death. I thought next I would try a father, and he came in dignity, feeling no doubt it was a challenge to him that he must meet as a man, and he talked in dignity until I began to tell him about his son's death. I had assumed that he knew some of the details of that death and, as a scientist, would care for other details that would help him participate in his son's last decisions, very thoughtful ones though tragic. As I mistakenly went on talking, his hands began to shake as if he had Parkinson's disease. He could not stop them, so there is no story, certainly no ending to a story, that can be found by communicating with the living who loved the young who are dead, at least none that I am qualified to pursue. A story at a minimum requires movement, and, with those who loved those who died, nothing has moved. It all stopped on August 5, 1949. So if there is more to this story for me to find, I shall have to find it somewhere else.

The silence, of course, could never be complete. Some things do remain—worn-out things, unconnected things, things not in the right place or clearly of another time: a worn-out fishing jacket, a few unrelated letters written by him, a few unrelated letters written to him, a childhood photograph that is hard to imagine as his, a picture in which something may be right.

A movie supposedly based on the Mann Gulch fire can still be seen now and then as a rerun on television. In 1952, while litigation was still in progress, Twentieth Century–Fox released *Red Skies of Montana,* filmed at the Smoke-jumper base in Missoula and on a fire just outside of town. The cast included Richard Widmark, Constance Smith, Jeffrey Hunter, and Richard Boone. At the beginning of the plot a foreman on a forest fire lights an escape fire, as Dodge did when the flames closed in; then he lies down in his own fire and the likeness to Dodge continues—his own men "do not heed him" and he alone survives only to live in disgrace.

At the end of the movie, though, there is another fire and the movie foreman again lights an escape fire. This time his crew heeds him, and everybody lives happily ever after.

Our story about the Mann Gulch fire obviously makes it hard on itself by trying to find its true ending. Here is this movie that lives on to rerun several times a year on TV and so has attained some kind of immortality by easily adding to small broken pieces of truth an old, worn-out literary convention. This added, life-giving plot is the old "disgraced officer's plot," the plot in which the military leader has disgraced himself before his men, either because his action has been misunderstood by them or because he displayed actual cowardice, and at the end the officer always meets the same situation again but this time heroically (usually as the result of the intervening influence of a good woman). By the way, this plot has often been attached to movies and stories about Custer Hill. Perhaps this is a reminder to keep open the possibility that there is no real ending in reality to the story of the Mann Gulch fire. If so, then let it be so—there's a lot of tragedy in the universe that has missing parts and comes to no conclusion, including probably the tragedy that awaits you and me.

———

This coming part of the story, then, is the quest to find the full story of the Mann Gulch fire, to find what of it was once known and was then scattered and buried, to discover the parts so far missing because fire science had not been able to explain the behavior of the blowup or the "escape fire," and to imagine the last moments of those who went to their crosses unseen and alone. In this quest we probably should not be altogether guided by the practice of the medieval knight in search of the holy grail who had no clear idea of what he was looking for or where he might find it and so wandered around jousting with other knights who didn't know what they were looking for until finally he discovered he was home again and not much different from what he had been when he started except for some bruises and a broken lance. I early picked up bruises in my wanderings, but I tried to shorten the length of the search by pursuing several quests concurrently. As early as 1976 I started the serious study of the Mann Gulch fire by trying to recover the official documents bearing on it and at the same time reacquainting myself with the actual ground on which the tragedy had occurred. For an opener I took a bruising boat trip down the Missouri River to Mann Gulch. Even earlier I had started with the archives in Missoula because, both as headquarters of Region One of the Forest Service and as the location of the base of the Smokejumpers sent to the fire, it was the center of the Forest Service's operations against the Mann Gulch fire. My opening jousts with both the archives and the ground went in favor of my opponents, whoever they were. My brother-in-law, Kenneth Burns, who was brought up within a few miles of Mann Gulch and was then living in Helena, said he would have no trouble borrowing a boat and taking us down the Missouri from Hilger Landing to the mouth of Meriwether Canyon where, as you'll remember, there is an almost vertical trail to the top of the ridge between Meriwether and Mann Gulch. Charles E. ("Mike") Hardy, research forester project leader at the Northern Forest Fire Laboratory in Missoula and only recently the author of a fine study of the beginnings of research in the Forest Service (*The Gisborne Era of Forest Fire Research: Legacy of a Pioneer,* 1983), had this early interested me in a scientific analysis of the fire and had been good enough to make the trip with us to give me a close-up view of his theories at work. Although we did not quite make it to Mann Gulch on this trip, I am grateful to Mike for starting me in the right direction.

A cloudburst was already waiting to challenge us at the top of the ridge. From the bottom of Meriwether Canyon we could both see and hear it making preparations for a joust with us. As we tried not to fall backwards to where we started in the canyon, we could hear the storm rumble and paw the ground. When we neared the top, it tried to beat us back by splintering shafts of lightning on gigantic rocks. There was a lone tree near the top, only one, and in case we had any foolish ideas of taking refuge under it a bolt of lightning took aim and split it apart; it went down as if it had been hit by a battle-ax. Trying to reach the rocks, we were held motionless and vertical in our tracks by the wind. Only when the wind lessened for a moment could we move—then we fell forward. With the lessening of the wind the rain became cold and even heavier and forced us to retreat from the battlefield on top. The rain fell on us like a fortified wall falling. By the time we reached the bottom of Meriwether, we were shivering and demoralized and my brother-in-law probably already had pneumonia.

All this was like a demonstration arranged to let us know that Mann Gulch had power over earth, air, and water, as well as fire. As the wind continued to lessen, the rain increased and fell straight down. It was solid now everywhere. It knocked out the motor in our borrowed boat, and we couldn't get it started again; after a while we didn't try anymore, and it took several hours to pole and paddle our way back to Hilger Landing. My brother-in-law was seriously sick before we got there; he would never go back to Mann Gulch. So for some time Mann Gulch was mine alone if I wanted it, and for some time I left it to the elements. I turned to the archives because I knew they would be dry and no wind would be there and the air would be the same air the stacks had been built around and nothing but a book or two had been moved since. The signs would demand "Silence" and even the silence would be musty, and for a time anything musty had an appeal.

The Forest Service archives in Missoula were about as hard to get anything out of as Mann Gulch. Although there wasn't much trouble gaining access to the Regional Library, there wasn't much stuff in its files on the Mann Gulch fire, and what was there was the ordinary stuff. Yet surprisingly even some of that was marked "Confidential."

I hadn't thought there would be much in the files, because you can't dip very far into the Mann Gulch story without becoming suspicious that efforts have been made to scatter and cover the tragedy. Besides, you are not far enough advanced in your thinking to do research on the Forest Service if you don't know ahead of time that the Forest Service is a fairly unhistorical outfit, sometimes even antihistorical. So when I first looked under "Mann Gulch Fire," the cupboard was practically bare, but before long I met the great woodsman W. R. ("Bud") Moore, who was then director of the Division of Aviation and Fire Management for Region One of the Forest Service. He is outspoken and devoted to the Forest Service and expects every American citizen, except the president of the United States, to be likewise. He sent out orders to round up all stray documents bearing on Mann Gulch that could be found in the region and make them available to me. They add up to a small but interesting file and contain most of the documents in my collection marked "Confidential." But all told, the Mann Gulch fire turned to ashes without depositing much in the offices of Region One of the United States Forest Service, and I had to make three visits to the Forest Service's Office of Information in Washington, D.C., before I had a good working collection of documents on the Mann Gulch fire.

It is harder to guess ahead of time how you will be received by those in charge of government documents than to guess what you will find in them. Ahead of time, I had guessed I would be sized up as a suspicious character up to no good: I was alone and peeking into government files and into Mann Gulch itself, which long since had been put out of sight and was better that way. Although Forest Service employees, I figured, would always be watching me with a fishy eye when I was around and even more so when I wasn't, there were not nearly as many spies as I had expected. They were mostly old-timers, and some of them had worked in the office long enough to know that some funny PR business had gone on at the time of the Mann Gulch fire. Most of the Forest Service employees who had a corner of an eye on me belonged to that element in most PR offices who are never important enough to be trusted with any of the organiza-tion's real secrets—they just know genetically that big orga-nizations have shady secrets (that's why they are big). Also genetically they like shady secrets and genetically they like to protect shady secrets but have none of their own. I gather that government organizations nearly always have this un-organized minority of Keepers of Unkept Secrets, and one of these, I was told, went so far as to write a letter to be read at a meeting of the staff of the regional forester reporting that I was making suspicious visits to Mann Gulch and reportedly and suspiciously arranging to bring back with me to Mann Gulch the two survivors of the fire. According to my source of information, after the letter was read the regional forester went right on with the business at hand as if nothing had interrupted him. And as far as I know, nothing had.

On the other hand, many of the men in the Forest Ser-vice whose main job is fire control are unhistorical for fairly good reasons. There have been millions of forest fires in the past; the Indians even set them in the autumn to improve the pasture next spring. What firefighters want to know is the fire danger rating for today, and as for me, I am not as important to them as the fuel moisture content for that afternoon. They are tough guys and I like them and get along with them, although I am careful about telling them stories of the olden days when at times it took a week or more to assemble a crew in Butte, transfer them to the end of the branch railroad line going up the Bitterroot, and then walk them forty or fifty miles across the Bitterroot divide to get them to a fire on the Selway River in Idaho. I tried to be careful and meek and not end any such stories with a general observation such as "Fires got very big then and were hard to fight." Firefighters prefer to believe that no one before them has ever been on a big fire.

Although it took more years than I had expected to get the information I wanted about the Mann Gulch fire, I know of only a very few instances where my difficulties were consciously made more difficult. To reassemble what was left of this fire, I needed the help of many more present members of the Forest Service than I can acknowledge in the course of this story, women as well as men. You must always remember the women, even if you are pursuing a forest fire, especially if you are pursuing it in a big institution. The new age for women had not yet worked its way through the walls of the Forest Service; still maybe it had a little. The women I worked with were in charge of the documents, the maps, and the photographs, and without them there would have been practically no illustrations in this book, or, for that matter, practically nothing to illustrate. Their attitude toward me was possibly a combination of women's traditional attitude toward men touched by an added breath of confidence in themselves coming from the new. They were certainly good and they knew it. When I entered their offices, whether in Washington or Missoula, they looked up and seemed to say, fusing two worlds, "Here's a man with a problem. What can we do to help him?"

Since I started to write this story I have seen women start taking over some of the toughest jobs in the Forest Service. I didn't believe I would ever see it, but now there are even a few women Smokejumpers. I can bear witness. One of them lives just a few blocks from me in Chicago. Her father is a faculty member of the University of Chicago—he is a distinguished statistician and one of the best amateur actors I have ever seen. She is a remarkable young woman—attractive, brainy, and tough. They tell me at the Missoula base for Smoke-jumpers where she is stationed that she is up there with the rest of her crew (which means men) in the training races run in full jumping gear or hard-hat equipment.

Several times in this story of the Mann Gulch fire I have tried to find places where it would be permissible to say that the story of finding the tragedy of the Mann Gulch fire has been different from the tragedy of the Mann Gulch fire. Tragedy is the most demanding of all literary forms. Tragedy never lets you get far away from tragedy, but I do not want you to think I spent ten years in sustained pain writing what I wanted to write about the Mann Gulch fire. A lot of good things happened along the way. Some things got better, and I met a lot of good people, some of them as good as they come.

It's hard to say when the pleasures and pains of writing start and end. They certainly start before writing does, and they seem to continue for some time afterwards. I met Bud Moore before I started to write, and he has become one of my closest friends. He and I soon discovered that both of us had worked in the Lochsa when we were boys and when the Lochsa was thought to be accessible only to the best men in the woods. Lewis and Clark had nearly starved there. Running into somebody who has worked in the Lochsa in the early part of the century is something like running into a buddy of yours who served on the battleship *Missouri* in

World War II at the time General MacArthur was on it. Those of us who worked on the Lochsa early in this century regard ourselves as set apart from other woodsmen and our other countrymen in general.

I had started writing this story before I met Laird Robinson, but was still heavy in research and was wandering around the Smokejumper base looking for any odd items I might have overlooked. Laird had been a foreman in the Smokejumpers and had injured himself landing on a fire and twice had tried a comeback but finally had to accept he was through as a jumper. He had been made a temporary guide at the Smokejumper base until they got him placed in a line of work that would lead him toward the top. He was in his early thirties and in the woods could do anything, and among other things he wanted to know more about the Mann Gulch fire. He put a high premium on friendship, and we soon were close friends and doing a lot of our digging into Mann Gulch together, as you will see from the course of this story. It is a great privilege to possess the friendship of a young man who is as good or better than you at what you intended to be when you were his age just before you changed directions—all the way from the woods to the classroom. It is as if old age fortuitously had enriched your life by letting you live two lives, the life you finally chose to live and a working copy of the one you started out to live.

I tried to be careful that our friendship did not endanger Laird professionally, and there were times when it seemed that it might. We were well along in our investigations when evidence appeared suggesting that Rumsey and Sallee had been persuaded by the Forest Service's investigator to change their testimony regarding the course of the fire at its critical stage. Persuading a witness to change his testimony to what he did not believe to be true was to me a lot more serious charge than scattering or burying documents that might bear on the threat of a lawsuit. So when I knew that I would have to try to find this investigator if he was still alive, I told Laird, "If you don't like the way this thing is headed now, just step off it before you get hurt. I can see it might hurt you, but there's no bravery in it for me because it can't hurt me."

Laird said to me, "Forget it. On my private list, friendship is highest." He said, "Anyway don't worry about me. The Forest Service and I can take care of ourselves."

So one of the pleasures of writing this story has been listening to the talk of first-class woodsmen, some old and some young.

9

For a long time, our story becomes the story of trying to find it, and like most stories of the woods this one must begin with the ground and with some questions to tell in which direction to look (since compasses only tell the directions, not which one to follow). To woodsmen, if you don't know the ground you are probably wrong about nearly everything else. To woodsmen, the ground often furnishes most of the questions and a good number of the answers, and, if you don't believe this, you and your story will most of the time be lost. A good woodsman who also was a fairly good story-teller would probably take only one good look at the crosses on the hill before the hill would be asking him, Why didn't the rest of the crew, after leaving their foreman behind, follow Rumsey and Sallee to the top of the ridge instead of taking off on a sidehill angle that was twice the distance the survivors needed to reach safety? Twice the distance on the ground says it doesn't make sense, and when something doesn't make sense to the ground the mind should be left with a question.

To try to keep up with Rumsey and Sallee is also to hear the ground ask questions all the time, and one in particular on which the story and eight lawsuits depend: Are Rumsey and Sallee going straight for the top, or is Dodge's fire driving them upgulch and so could be the fire that prevented the sidehilling crew below from escaping, as Thol Senior was to charge? If Rumsey and Sallee ran alongside Dodge's fire straight to the top, another question follows: How could this "escape fire" burn straight up the slope and across the path of the main fire, which was being driven upgulch by a strong wind? Finally, the crevice itself through which Rumsey and Sallee crawled to safety forever asks its big question: What did Diettert see in me or the ground beyond that he did not like and so did not crawl through me to safety? Without this question the story of the Mann Gulch fire would lose one of its most moving parts. Diettert was a fine jumper and a young scientist with unusual promise; yet he chose not to go through the crevice and died about 275 yards from where he left Rumsey and Sallee.

When Laird and I first started investigating these questions, we assumed as almost self-evident that a few moments after the survivors crawled through the crevice, a fire must have pushed upgulch and closed it, and that for a fatal distance beyond there must have been no other opening in the reef. We had been trying to remember the reef directly above the crosses as something like a Wall of China with only one breach in it through which the survivors escaped and, beyond that, an impregnable cliff. We should have known better, but such is the power of theory over rocks that it can make rocks into solid cliffs, which, however, when looked at close at hand, present openings wide enough to drive hay wagons through. On our 1977 trip to Mann Gulch, Laird had discovered that only occasionally was the reef solid enough to keep a fast climber from crawling through or over it.

Suddenly we felt the need for something we had needed for a long time without recognizing it—to get the two living survivors of the fire back into Mann Gulch with us. Suddenly we also realized that we probably didn't know about a lot of things we thought we did but maybe only dead men knew. Survivors after nearly thirty years sound unreal enough to be dead, and, as far as we knew at the time, one or both of them might be. Nobody even at the Smokejumper base in

Missoula knew whether they were legendary but alive or just burned-out flames burnished with legend.

———

We knew, of course, that of the three survivors of the fire, Wag Dodge had died soon after (in 1955), but we had a hard time finding out whether Walter Rumsey and Robert Sallee had addresses on this earth. You will discover, if you ever try to find out about a mass tragedy, that people believe the few who survive die soon after it. There is always that strong mental connection between a first-class catastrophe and the "kiss of death," and, in the case of the Mann Gulch fire, it seems there had been such a lasting kiss, since not only Dodge but the pilot, Kenneth Huber, were dead soon after the fire, as if they were also its victims. When I found that nobody at the Smokejumper base was sure whether Rumsey and Sallee were alive, I began to think of a poem by Sandburg about a "little fliv of a woman" who wrote a letter to God but it went to the Dead Letter Office, "where all letters go addressed to God and no house number."

Fortunately, the basic tools of scholarship are much the same the world over, whether they are used under the ever-ready pigeons on the edge of the roof of the British Museum or in the presence of the white mountain goats that flit among the Gates of the Mountains. Wherever, you had better soon start looking for "first-hand sources" and, in order to discover what they are and where you can find them, you had better be good at "bibliography." Scholars of the woods know that one of the best bibliographical reference works to consult is the postmistress of a nearby logging town. An ex-postmistress at my second home of Seeley Lake, Montana, who is a sort of yellow-pages directory of the loggers of the Northwest, told me she knew of a Sallee who was related to another Sallee and this other Sallee might be the Sallee I was looking for—if so, he was working in a sawmill west of Missoula toward Frenchtown.

There is a Hoerner Waldorf—Champion paper mill out there, and it was going full blast when I arrived. The outside, at least, looked just like the mills I had worked in when I was too young to work in a mill. I couldn't find anybody in the offices, least of all in the personnel office, where there was a sign saying as always that they didn't need any more help. Only one of all these offices was occupied; it said "The Nurse." So the composition of a successful sawmill must still be the same as when I was first a millhand and was told I would never be an honest-to-God millhand until I lost a couple of fingers in the saws. The truth on which this ancient adage rests is that a sawmill is a large building full of moving chains, belts, and saws, all the chains and belts pulling toward saws, so if you or your clothes get caught in a chain or a belt, you know where you are going—you are going to the nurse. The composition, then, of a successful sawmill is a sign saying "No Help Wanted," all the wanted help inside the big building working close to belts and saws, and the only office occupied by one nurse who can sew on fingers.

I felt encouraged. The nurse looked and sounded as if she were a French-Canadian from Frenchtown. She said, "What can I do for you?" and I replied, "Do you know whether a guy by the name of Sallee ever worked here?"

She asked, "Which Sallee? If you shake all the pine trees between here and Frenchtown—maybe between here and Superior—a lot of Sallees will drop out. Either a Sallee or a Deschamps or a LaCasse. What's the name of your Sallee?"

I said, "Robert."

"Isn't that funny!" she said. "I have some friends by the name of Cyr who just got back from a vacation on the West

Coast, and they stopped and visited Robert Sallee in Portland. If you'll wait just a moment, I'll telephone the Cyrs and get his address."

So some lessons about the bibliography of the woods are fairly simple, such as the one about the postmistress. It is also simple if you are looking for a French-Canadian in the woods—all you have to do is find another one.

I flew to Portland to have a talk with Sallee, and he told me that he was sure Rumsey was alive, although he hadn't seen him in years. When he last saw Rumsey, he was in charge of some kind of soil conservation work and lived somewhere in the Southwest, but Sallee couldn't remember where, although he thought he had the address at home. The address he later sent me was not in the Southwest but in Boise, Idaho, and a month or two later Rumsey answered me from Lincoln, Nebraska, where he and my letter had been transferred.

When first seen in person, the two survivors were unexpectedly real, and it is surprising to find that ghosts are real. They seemed big men for ghosts; both had become very successful in their professions, and it showed; both remained professional outdoorsmen, and that also showed. Sallee has stayed with timber and the mountains and works for Sandwell International, a consulting engineering firm. Rumsey returned to the plains from which he came and specialized in irrigation and soil conservation; he was to be killed in an airplane crash in 1981.

It didn't take long after meeting them to discover they depended upon being curious. Among other things, they were curious about me. They couldn't quite figure "what I was up to" and "what my game was," and it took a winter of letter writing to make myself seem real to them and on the up-and-up. But it worked both ways. Laird and I were curious about them, as you would have been. I wanted to see them in the crack of the earth through which they had crossed from death. Since both of them had told me they had spent much of their afterlife trying to forget the fire that they alone could remember, I also became interested in seeing what they did and didn't remember. I thought, just as an intellectual exercise, it would be interesting to observe what real ghosts remember of the death they did not die but those only seconds behind them did. And, of course, it would be moving to see two real ghosts together again who had been roommates in the first life and so had helped each other to a second life. I was not surprised to find that my chances of getting them back into Mann Gulch depended upon their being curious about the same things I was, which they were. Out of curiosity, then, all four of us agreed to spend the day of July 1, 1978, in Mann Gulch.

The shortest possible version of the long story of finally getting together in Mann Gulch is that finally we did. Even this shortened version should include the detail that Laird's boat wouldn't work after he had dragged it over the Continental Divide from Missoula to Helena the evening before. But he remembered a hunting pal who lived twenty-nine miles out of Helena on the Missouri River and had a jet-propulsion boat, the kind that can be landed in shallow water, and you can bet there are no docks where we were going to land. At three o'clock in the morning, Laird returned to Helena with the borrowed boat. We gave him a couple of hours in bed before pushing the boat into the Missouri and heading for the Gates of the Mountains.

———

We landed the big boat at the mouth of Rescue Gulch, which must be among the earth's special gulches to Rumsey and Sallee. When they crawled through the reef out of Mann Gulch, they crossed into Rescue Gulch, and it is near the head of Rescue Gulch that they found the rock slide in which they dodged from one side to the other as the main fire flapped by them. It is the head of this gulch that Hellman reached after the fire caught him crossing out of Mann Gulch, and it is up this gulch that Jansson and Sallee led the rescue crew the night of the fire and met Rumsey coming down trying to reach the Missouri at midnight for a canteen of water for Hellman.

Approaching Mann Gulch from Rescue Gulch is approaching Mann Gulch from the side, and Mann Gulch can't be seen until you look down into it after reaching the top of the ridge. However, if you know where to look from the mouth of Rescue Gulch, you can see Hellman's cross close to the top. It is up this gulch that Jansson took Gisborne, and Jansson, having been head of the rescue crew, knew where Hellman's cross was and from the mouth of Rescue Gulch pointed it out to Gisborne, who took two hours getting there, stopping every hundred yards by prior agreement. Probably it was because Gisborne died of a heart attack on the way out that the others with me insisted I not try to make the climb, being twenty years older than Gisborne was when he died and, like him, having had heart problems. They even said they had been told in Missoula not to let me go. Finally, I had to get personal and tell them, "Look, there is a mountain downriver no farther than twelve miles from here by air that also looks over the Missouri. It was named by my wife when she was still a girl, and she named it Mount Jessie after herself, although she lived an otherwise modest life. At her request her ashes are there now. Nobody should feel bad if I should remain behind on one of these hills that looks her way." Sallee reached over and took my pack off my shoulders, and we started climbing.

He and Rumsey walked ahead toward the top of the hill. They seemed to get bigger instead of smaller as they climbed. Sallee, who at seventeen had been underage for the Smokejumpers, remained almost oversized. The afternoon I talked with him in Portland, in the board room of Sandwell International, he looked big just sitting there. He answered questions suddenly, especially if he didn't like them. Almost as soon as we sat down, I asked, "Is it true you lied about your age to get into the Smokejumpers?" He never moved from his elbows. "Who told you that?" he asked. When I said, "Somebody from Frenchtown," he said, "Yes."

Not long afterwards I asked him how he accounted for the fact that he and Rumsey, the two youngest and most inexperienced members of the crew, were the only ones to survive. This brought him off his elbows. "What do you mean, the most inexperienced? What difference does it make that we were in our first year as Smokejumpers? Jumping had nothing to do with what happened in Mann Gulch. Mann Gulch was nothing but a footrace with a fire. I was brought up in the backwoods in northern Idaho, where I had to go four miles each way to school, and I ran it. I was the best walker in every crew I worked on, and I made a point of showing I was, because it showed I wasn't underage. Rumsey was from the plains and from a small town, and he was tough. We were roommates, and, if things didn't go right, we saw that we never got far apart."

Probably one reason Sallee has done so well in business is that he doesn't fool around with questions. He himself attributed his position at Sandwell to his absolutely white hair. Probably both reasons are about the same.

Rumsey was a lanky Kansan and liked to have time to think and to remain in doubt about quite a few things, partly because, as a Kansan, he found it uneconomical, even dangerous, to think of too many things at once. He was the one who thought only of "the top of the ridge, the top of the ridge."

Walking behind, Laird and I could see the two in the same picture frame and, high above in the same picture, the cross of Hellman. We should like to have had a photograph of the two climbing toward the cross, but the cross was too far away to have shown—only in our memories are all three in the same frame, two returning and one forever there. We kept wondering what they were talking about, although we made a point of staying far enough behind so we could not hear.

Then they stopped and waited for us to catch up. Sallee said, "Hellman's cross is not in the right place." Rumsey said, "I agree." The cross was still a half mile above, and the grass everywhere was a foot or more high. I said, "We're lucky we can see the cross from here," but I didn't argue. It was Rumsey who had started down Rescue Gulch at midnight to get water from the Missouri for Hellman, and it was Sallee whom he had met halfway up leading Jansson, the doctors, and the rescue crew. The two together had had no trouble finding Hellman in that land which in the dark had lost its identity. Twenty-nine years later they were going to give themselves a test to determine whether, unlike most mortals, they could find their way in this world and the world that died behind them.

Sallee said, "The cross is too close to the rock slide where the fire went around us." He said to Rumsey, "Remember, we yelled after the fire went by, and Hellman answered about thirty yards away."

Rumsey said, "I don't think his cross is even on the right side of that draw."

Sallee said, "I'm positive," a word combination he likes. "On our way back this afternoon, let's go by the rock slide and check."

Later that afternoon, when we came back and were looking down toward the cross but were still quite a way from it,
Rumsey said, "I think we are right. It should be farther from
the rock slide, on the other side of the draw and lower down.
His answer came from below."

Sallee said, "I know positively it is wrong. There was a big,
flat rock lower down on the other side of the draw, and we
put him on it to keep his burns out of the ashes."

"That's it," they said when we got close to the flat rock.
"His cross should be here."

Sallee said to Rumsey, "We could be doubly positive if we
could find the tin can Dodge left you when he and I started
for the river and help. Remember," he said, "he left you a
can of Irish white potatoes and his canteen. He had thrown
everything else away."

Rumsey looked behind him and said, as a Methodist, "By
gosh, there's a rusty old can."

I started to reach for it, but Rumsey stopped me. "Don't
touch it. Let me think for a minute." Then he said, "Of course,
I didn't have a can opener with me—only my jackknife. Besides, Hellman didn't want the potatoes, just the juice, even
though it was salty and would make him more thirsty, so I
jabbed my knife twice into opposite sides of the top of the
can, one of them to let in air for the juice to come out the
other one." Then Sallee reached down and handed Rumsey
the rusty can, and it had two knife jabs on opposite sides of
its top. They had passed their own test.

So Hellman's cross is not properly placed, as two ghosts
who are woodsmen could tell a half mile away. It should be
over thirty yards to the west of the rock slide, across a draw,
lower down the hill, and next to a big, flat rock.

The night before in Helena, while Laird was down on the Missouri somewhere trying to find a very special kind of friend who owned a special kind of boat with a motor that worked, Rumsey, Sallee, and I sat several hours around the dining table before carrying out the dishes. We were sure that tomorrow there would be four men in Mann Gulch who collectively knew more about its tragedy than would ever be assembled there again. Two were the only living survivors, both of them still outdoorsmen; another was one of the Smokejumpers' finest modern foremen, who had become information officer at the Smokejumper base in Missoula, answering all questions the public could think of asking about getting to a fire and back again; and I had been on some big fires too, and, if that was before the other three were born and made me a little slow of foot, I compensated by having collected the best existing file of documents on Mann Gulch, including statements Rumsey and Sallee had made soon after the fire, which I intended to carry in a packsack with me into Mann Gulch the next day when we convened as a complete court upon ourselves—plaintiffs, defendants, witnesses, attorneys, judge, and jurors. In such distinguished company, including the packsack, it would be hard for any one of us to remain far wrong. At least, not for long.

Since it takes a lot of daytime just to get in and out of Mann Gulch, we were selecting our targets in advance so we wouldn't end up tomorrow as scattered over Mann Gulch as the cargo had been after it was dropped from the plane. Inevitably, our sights all lined up on the same target, the scene of the catastrophe and the crosses.

You would have picked the place of the crosses yourself; nearly anyone would, so ancient and binding are the connections between drama, religion, and the top of a hill. The Christian scene of suffering, where hill meets sky, has been painted so many thousands of times that something within it must direct it to paint itself.

On pages 10–11 of the photo gallery in this book is a photograph in which much of what was left of the catastrophe of Mann Gulch appears in quiet composition but is short of being classic in the composition of catastrophe. I found it in a Forest Service file and so for a moment felt that nature had composed it, only later realizing that it is the work of a partly informed photographer with a good enough eye to combine the standard principles of photographic design with much of the ground on which the historical and dramatic catastrophe happened. But it is short of a perfect combination of art and history because an error in photographic art seriously divests it of the intensity of the real thing, and, what is more, it omits a slice of adjoining ground which would not have been hard to include and on which two of the most important events in the tragic story occurred. Still, it was the best of the historical photographs of the scene of suffering I had found, so it was the photograph Laird and I had to work with as we started our quest for the missing parts, and you too should start with it.

Despite its faults, this composition of the scene effectively observes the traditional three-part division of foreground, middle distance, and something at the top to frame it, each of these topographical and aesthetic formations making visible a separate dramatic and historical part of the tragedy, with each given its proper size.

The foreground is the darkened land of death extending completely across the bottom of the picture—dead trees, burned, fallen, and rotten, a broken stub of one of the fallen trees standing close to and as high as the cross to expose its death. A thistle, the only thing living, fronts the white cross of Stanley J. Reba, which is the lowest on the hill of all the crosses and which possesses all the rest of the photograph and says what the rest is all about. Topographically, historically, and dramatically the end of this tragedy properly rests upon this dark foreground.

The upper frame of the tragedy is also composed of topography, history, and drama. The reef of rocks at the top was probable salvation if it could be reached. Near the top is the highest standing tree on the slope (marked X on the photograph). It comes out of the middle distance, but its top connects visually with the reef and even the sky, suggestively just to the left of an opening in the reef that might be the crevice with a juniper bush on the other side. Likewise, the dead tree might be the tree Dodge stood beside when he lit his escape fire. Above the frame is one faint semi-arid cloud, perhaps a reminder that Smokejumpers sooner or later return to the sky from where they are dropped, certainly a sign that rain will not come for days.

The seemingly anonymous middle distance between Reba's cross and the reef is where nearly all the tragedy occurred—anonymous since no one who lived saw what finally happened there, anonymous also since even those who died there didn't see much of what was happening to them because at the very end there was only heat too hot to breathe and not enough oxygen to keep the brain alive. A photographic mistake, however, obscures one cause of the fatality of this middle distance—the steepness of the slope between the foreground and the reef. It is about a 76 percent slope, meaning that for every 10 feet you walk forward, you also gain 7.6 feet in elevation, climbing at about a 45-degree angle. But in the photograph, instead of the middle distance rearing up like half a cliff, it looks more like the gentle, grassy slope of Custer Hill. If the photographer had had the right equipment with him, he could still have featured Reba's cross in the foreground without flattening the scene of suffering behind it. The more the photograph erroneously reduces the grade of that slope, the more it erroneously reduces the actual speed and intensity of the fire that went up it, even the length of its flame; the more it erroneously flattens the scene, the more it eliminates the emotional realization of young men that it will be impossible for them to climb the slope as fast as fire.

Since the four of us at one time or another had worked on mapping crews, we knew that to trace the movement of men or fire precisely we had to start as we would in mapping the course of a stream—at a point already located with certainty on the ground and map. To begin, we would try to find a nearby "location point" established by an outfit which always seemed to know where it was in the woods, the United States Geological Survey. The location point usually would be a lead plug driven into the ground at the top of a prominent peak and stamped with USGS and the elevation. Then, after we had compass-and-paced the drainage system from there, we would continue our line until we could connect with another established location point. Unless we began and ended with known points, we might have accurately traced all the bends in a stream, but how the hell did we know where the stream was?

We agreed without argument that we needed to locate two key points on the hillside to map the final movements of those who died there. For sentimental and theological reasons alone, the first point had to be the opening in the reef through which Rumsey and Sallee crossed out of Mann Gulch. Late in the afternoon of August 5, 1949, it was the opening to salvation; now it was the point, like magnetic north, that had beckoned the two survivors back to Mann Gulch—their primary reason to return to the scene of the fire was to pass through the crevice again.

Second, and most important of all, was to locate accurately the spot where Dodge had set his fire. A missing part of the story, threatening to remain forever missing, was the tragic ending itself, whatever it was that happened after Dodge's men passed this spot. No one lived who saw this missing part, and a storyteller who wanted to find it would have to know that the missing part of his story pivoted about a single point. There couldn't even be the cartography of a story without an accurate location of the spot where the escape fire started.

It may seem not worth mentioning again now, but the rest of this sentence is placed here with some care—Dodge said that from his fire to where he found Sylvia still alive was about 150 to 200 feet east (upgulch) and about 100 feet below him.

Also, I had a high priority of my own, the cross of Henry Thol, Jr. It was the cross farthest upgulch and closest to the top of the ridge—his cross would define an important outer boundary of the tragedy. In addition to its being useful to the cartography of tragedy, there was this personal reason: I find in trying to record the tragedy of a good many characters who were young and much alike that a few remained distant from me and anonymous and were always dead—only some came close to me and asked me to visit their crosses when I returned to Mann Gulch and to try to be of some comfort to them. Thol's cross was one I always visited, and when I did I would try to imagine a little of what it must have been like to be highest on the hill and not quite high enough. I also wished I could have been of some comfort to his father, but he was so enraged in his grief I probably couldn't have been, even if I had had the chance. Yet, even if Thol had to die, his father should have found room to be proud of where his son is, the nearest to the top.

When all this is added up, it comes close to a day's work in Mann Gulch, given the amount of energy it takes just to move around in Mann Gulch after getting there. But in addition, Laird and I had a piece of unfinished business we hoped to wind up on this trip, although we hadn't mentioned it to Rumsey and Sallee for fear of suggesting what we wanted them to say. We still had only a guess for an answer to the question that had first sent us on our quest to find the two survivors and persuade them to come back to Mann Gulch with us: Why didn't the rest of the crew, only a few moments behind you, follow you straight to the top of the ridge and safety?

Despite a long after-dinner session, it wasn't possible to think of everything, but we felt we had a fairly good plan. We couldn't tell what part of it we would be starting with, since we planned to approach Mann Gulch by way of Rescue Gulch and so would have to guess just where we would come out on top of the ridge. There was no question, though, that there was just one place of all places that Rumsey and Sallee wanted to see first when they reached the top and looked down—the opening in the reef through which they had crawled out of Mann Gulch. We all hoped that when we came to the top we would be close to it.

———

Near the top of the ridge, we ran into a deep game trail sidehilling toward what would probably be an open, wide saddle—we followed it because the footing was good, not because we were looking for an open, wide saddle. Even before we reached it, Sallee said, "It's the same God damn pass the Forest Service investigator tried to make me believe was where Rumsey and I crawled out of Mann Gulch."

Over a quarter of a century later, Sallee was still angry

because some company detective had tried to make him believe that he didn't know where he was at the big moment in his life.

French Canadians don't take kindly to being told they don't know where they are in the woods. Sallee said, "He wouldn't even stop for a few minutes to let me go up the ridge and show him the right crossing."

The investigator's pass was certainly a low and wide one—the top of the reef had decomposed and completely disappeared for thirty or forty yards. "It's a good-looking pass," I told him. "A couple of threshing outfits could cross here abreast."

"That's one of the big troubles with it," Sallee said. "Rumsey and I had to squeeze through a crevice in the reef one at a time."

I knew without looking at the documents in my packsack (which Sallee still carried) that it had been a close squeeze at the top for the survivors. Both of them early had testified that they first saw the top of the ridge as they approached Dodge lighting his fire; Rumsey stated that when they left Dodge they followed the upgulch side of Dodge's fire until they spotted on their right a pile of rocks at the crest and headed for it; they have always agreed that Sallee was the first to edge through the crevice and Diettert did not follow. I asked Sallee, "Why didn't you go straight uphill from Dodge's fire to this wide pass instead of angling upgulch under a reef that might not have another opening for a long way?"

He answered, "Because by then there was a fire in this pass."

So there at long last was the answer to one of the questions that the survivors alone could give. In their testimony they had sometimes insisted that, after leaving Dodge at his fire, they had gone straight for the top and at other times seemed less sure. Their answer is that they would like to have gone straight for the top, and did for a way, but a fire on the top of the ridge had closed a wide saddle above them and forced them to angle to their right to the next opening upgulch in the rocks above them. The sight of fire or smoke at the top of the ridge may also explain why Diettert turned aside at the crevice and why most of the crew kept angling upgulch. But Sallee's answer raises a new question: What fire was it, anyway, that was running upgulch on top of the ridge, cutting off escape from below? Was it a branch of the main fire that had been chasing them up the gulch? Or was it their foreman's escape fire, turning upgulch after reaching the top of the ridge?

We didn't sit down to rest from the climb. We just kept following Sallee upgulch on the ridge looking for the crevice. For one thing, it was too cool in the winds up there to sit in one's perspiration, although it was the first of July and the middle of the morning. The world of high altitudes is hard to believe from below. It is never what flat-country people call normal. That first summer I tried to reach Mann Gulch, a cloudburst turned our perspiration to ice and ended my brother-in-law's friendly visits to Mann Gulch with a case of pneumonia. Most of the times I have been there, though, it has been so hot in the day that rattlesnakes stay in the shade of their holes, although even then the cool is never far away. The heat of the mountains is not the heat of the plains, which stay hot all night. Remember, on the day of the fire it had been a record ninety-seven degrees in Helena in the shade and it had to have been a lot hotter than that in the burning gulch, but the night of the fire it had been so cool near where we were then standing that the rescue crew had given Hellman their one blanket and had huddled close to Sylvia to keep him warm. As we climbed up from the river, we soon left summer behind and were walking through the world of spring flowers, beautiful blues and yellows, lupines and vetches, and balsam roots looking with wide brown eyes at ghosts and intruders. But it became even cooler toward the end of the day, so we never sat down, tired as we were, until we got back to the boat and water level.

It took Sallee and Rumsey only a few minutes to announce their crevice. Within twenty or twenty-five yards upgulch from the open pass, the reef appears again and for fifty yards or so is solid cliff—and then there is a crevice. The two survivors took turns crawling through it, trying it on for size, as it were, and seeming to find that the passing years had shrunken it and made it a tighter fit.

I suppose they were amusing because of the difficulty they were having fitting through the crevice. But it was not possible to watch them without mixed feelings. For me, it was hard to look at the crevice and not think of Diettert turning away from it. It must have been even harder for Laird. He came over to where I was standing and finally said, "I wouldn't have gone through there either. You couldn't tell from here, especially in the smoke, whether you would only be walking into a firetrap on the other side."

"In the smoke," I said, "nothing could have looked sure."

"I would have gone with Diettert," he said.

There was no more to that conversation. Where could it go?

When it came Rumsey's turn to crawl through the crevice, I followed him. Then we went looking for the juniper bush he had fallen into, so tired he had almost let matters rest there.

"That's it, that's it!" he said. "I was exhausted; I almost stayed there." But when we got closer, he began to shake his head until words finally came out of it. "I just can't believe it. I just can't believe I have been wrong all these years.

It isn't a juniper bush. It's a dwarf, distorted alpine pine."

After a while, he repeated, "I just can't believe it. You just can't believe anything that happens in a forest fire."

He didn't realize at the moment what a great amount of truth he had uttered, especially about the juniper bush.

———

A day is an inadequate piece of time in which to get anything done in rough country, so we had split the job of locating the two remaining points of the triangle—where Dodge had lit his escape fire and where Thol had fallen. Naturally Rumsey and Sallee, who had seen Dodge light his fire, wanted the job of finding where they had cut loose for the top, and Laird went with them, having himself made the escape fire something of a specialty. At the time, I was glad to have the assignment of visiting Thol's cross and, on the way, of passing Diettert's. There are moments of the fire that I have tried to live through again and again with Thol and Diettert— especially the moment when they came nearest to escaping but realized they were not going to make it.

To Thol's father, the retired ranger, his boy was always a boy. When someone on the Board of Review testified that the crew consisted of highly experienced firefighters, his grief and anger again swirled up out of his son's ashes. "My boy spent two years in the Condon [Ranger] District as fire guard. I knew him—I knew him well—this I say of my own boy—he was a long way from an experienced firefighter. He could barely handle tools. He could handle tools well, but as being productive in handling tools, well he was not—and most of those boys—pretty much the same."

This outcry is even more confused than its sentence structure. The young members of the crew couldn't have been old-time experts with tools, but they had to be pretty good with the few basic tools of firefighting, and the son of a ranger had to be damn good. Anyway, what did fire tools have to do with the fate of the crew in Mann Gulch? They died in fire but had no chance to fight it. All the crew did with their tools was throw them away on orders from their foreman. The history of the Mann Gulch fire is the report of a race with fire to death. In this race, the cross nearly as close to the top of the hill as Thol's is Navon's, and Navon was twenty-eight; he had been a paratrooper for five years and a lieutenant at Bastogne. What do you expect of your nineteen-year-old son who had only eight jumps on a fire to his credit but was even a little closer to the top than Navon?

The crosses are simple and impressive and were made by the Smokejumpers in Missoula. As the photograph of Reba's cross shows, they are concrete with the name on a horizontal bronze plate. That's all there is, and that leaves everything very lonely up there, where no one goes walking or visiting. Besides rattlesnakes, there are mountain sheep and mountain goats, which also appear lonely, and, as the photograph shows, the fire left few living trees standing, so there is not even the friendship of shade.

Actually I had come to Thol's cross to do two jobs: besides locating Thol's cross accurately, I was to inspect the condition of his cross and that of several others nearby. Laird and I had promised Edward Heilman, director of aviation and fire management, that while in Mann Gulch we would inspect all the crosses and let him know if any needed repairing or replacing. To be sure that promise was kept, the four of us on the hill on July 1, 1978, had divided the crosses among us.

Thol's cross is straight underneath an open saddle on the ridge where the surface of the reef had decomposed, so the reef did not act as a barrier to prevent him from reaching the top of the ridge. It is impossible to say exactly how close

Thol's cross is to the top because without instruments it is impossible to locate the crest on the wide, smooth surface of the ridge. We had recently found a 1952 Forest Service contour map of "Part of Mann Gulch Fire Area" (see pages 8–9 of the photo gallery) where the distance from Thol's cross to the top of the ridge is probably less than 100 yards. By the same map it is 140 yards from his cross to Diettert's. The distance from his cross to the place where Dodge started his fire will vary according to where you think Dodge lit his gofer match. If you accept the two places as correctly located on the 1952 contour map, the distance is 390 yards.

It is not certain that Thol's cross marks precisely where Thol fell. I found his Forest Service flashlight with the batteries still in it seven or eight yards above his cross, and many of the crosses must be downhill from where the bodies fell. On that 76 percent slope of crumbled cliff and dry, slick grass it is difficult just to crawl without slipping; those running and stumbling, exhausted and terrified, may have rolled downhill after they fell. Remember that Jansson had another theory, even more terrifying. Noting that nearly all remaining non-flammable objects (such as watches and wallets) were found uphill from the bodies (unless found under a fallen body), he theorized that the power of the fire was so great it carried these objects with it. Both theories are probably correct.

———

On my way back from Thol's cross, I stopped to check on the condition of the crosses near his, all the time keeping an eye on the survivors, who were supposed to be locating the spot where Dodge lit his fire. At a distance they seemed stylized, as if they had been assigned to a repetitive one-act play with three repetitive movements—they climbed back up the hill to the crevice, gestured there to the sky with seemingly no accompanying words and no response from the sky, and then retired downhill to gesture again, here seemingly just to themselves. Such vast country that swallows up words also makes gestures seem very small. It makes humans themselves small, two- and three-foot miniatures. At a distance they were animated gestures that had found a script in the rocks and were repeating it until they could get it right. The identity of the animations could be told only from the gestures. The biggest gestures were Sallee, being French. Rumsey was the Methodist gesture with one hand raised to the sky. You could guess what the gestures meant by seeing where Laird went afterwards.

Then I noticed that, when the gestures and the accompaniment returned downhill, they did not always return to the same place. That was the tip-off—maybe there was an argument going on between Sallee and Rumsey about where Dodge had lit his fire. It certainly wasn't what it was like at the crevice, through which they had crawled back and forth with difficulty but in complete agreement.

And, sure enough, they were arguing about where they had left their foreman behind. But as the words paralleling their gestures became recognizable, it became clear they shared basic agreements. Most important was their certainty that they had located the opening in the reef through which they had edged their way to safety. In the coming conversation note how the crevice is the North Star that guides all arguments and tells the directions of all movements and measures their distances.

The second basic agreement was their certainty that a Forest Service map prepared a month after the fire ("Helena National Forest, Mann Gulch Fire–No. 35, August 5, 1949: Fire Progress Map") incorrectly locates the origin of Dodge's fire, a certainty essentially derived from their first one—the map's location of Dodge's fire isn't compatible with the location of the crevice. I had long had some misgivings about Map No. 35 because it had been prepared immediately after the fire, so I had gone in search of another and had found the 1952 contour map, which I had come increasingly to respect. I should add here that on the 1952 contour map the origin of the escape fire is located close to where it is on Map No. 35.

I joined the circle of gestures just as Sallee was saying, "The map places it too far back toward the mouth of the gulch. If that's where Dodge lay down in his escape fire, then I couldn't have seen him entering his fire or the crew angling up the hillside after passing him. There is a lateral finger ridge here that would have been between us and would have blocked my view from the crevice."

Sallee's argument will be easier to follow if visualized on the photograph of Reba's cross. Sallee has just decided that the origin of Dodge's fire was at the base of the highest dead standing tree (marked X) between Reba's cross and the reef, so close to the reef in fact that it is only sixty or seventy yards below it. Their crevice, they had decided almost immediately, is an opening in the reef just slightly to the right of the top of this highest tree. Remember, Sallee has always been positive that, coming out of the timber, he looked ahead and saw Dodge lighting his fire near a lone standing tree. In this photograph, then, it is almost as straight as a ruler from Reba's cross to the two most important points we were trying to locate—first, from Reba's cross to the origin of Dodge's fire as located by Sallee (X) and second, from there to the crevice both Sallee and Rumsey had just crawled through. Sallee accordingly is repudiating both Map No. 35 and the 1952 contour map, which locate Dodge's fire outside the photograph, slightly farther downgulch than Reba's cross.

It makes a distinct dramatic difference to this photograph whether Sallee or the maps are right. If Sallee is right, then this photograph shows the whole final tragic scene, both where the final tragic decisions were made (at Dodge's fire) and the adjoining field of final suffering. If, however, the maps are correct in locating the origin of the foreman's fire outside the upper left corner of this photograph, then this photograph takes in most of the field of suffering but not the point where the big decisions were made.

To follow Sallee's argument against the location of the origin of Dodge's fire on the maps, we must focus on the lateral ridges running down almost at right angles from the reef. These elevations have been formed by the dry gulches running on each side of them, and though these lateral gulches have water in them usually only in the spring, millennia of springs have cut them deep, a lot deeper than is suggested by this photograph. Sallee is correct that one of these lateral ridges (seen in the upper left corner of the photograph) would have cut off the view from his crevice to where the maps locate the origin of Dodge's fire.

Clearly, to Sallee the number one condition a map of the Mann Gulch fire must meet is that the origin of the escape fire can be seen from where he took his last look at Mann Gulch, a cartographic condition to be respected by anyone who has learned from the Bible to respect the last look behind.

"Besides," said Rumsey, "the map's location of the origin of Dodge's fire makes where we started just too far from the crevice. We never could have made it to the reef if we had to go that far."

"What's more," Sallee added, "it was beside a tree. I remember Dodge was stooped over near a tree starting a fire with his cigarette lighter."

"It was a gofer match," I said, because I had put myself

to some trouble to find out what a gofer match is—a paper match pulled from a matchbook. They were so unreliable that a firefighter trying to light a backfire with one would usually have to go for a second match.

"I don't remember a tree at all," Rumsey said.

"How could you forget?" Sallee asked. "There it is, still standing," and he pointed at the dead tree in the photograph that is nearest the top of the ridge.

Rumsey, with the composure of Kansas, repeated, "Dodge wasn't standing near a tree."

On the hill, the base of the highest standing tree is in a much deeper depression than it appears to be in the photograph, and Rumsey believed that Dodge lit his fire not in a depression behind a finger ridge but near its crest and even closer to the reef than Sallee places it. So that is how matters stood, even after we called it a day and started back for the river. The difference between the two of them does not appear to be great—only forty or fifty yards—but it could be important.

It was on our way back to the river that we stopped at the rock slide whereupon they collaborated in relocating the site of Hellman's cross. It was a brilliant job, and Plato, who talks as if all knowledge depends upon memory, would have been pleased to see them performing this cartographic feat of relocating Hellman's cross by remembering a flat rock and a tin can with two holes in the top of it.

At the mouth of the gulch, a strange boat and two strange men were waiting for us. The men became more familiar as we got closer and soon agreed to help us drink the beer we had brought along to save ourselves from dehydration. They turned out to be the boatman who conducted tours through the Gates of the Mountains and a friend of his whom he had brought along to see the two living survivors of the Mann

Gulch fire in the flesh. The standard tourist excursion starts at Hilger Landing and goes as far downriver as the mouth of Mann Gulch, where the tourists get a lecture on the Mann Gulch fire. The man who gives the lecture had come all the way downriver and had waited for us at the mouth of Mann Gulch in order to see his lecture alive. He had his friend pose Rumsey and Sallee for various photographs and was careful to include nothing that was not in his lecture, particularly not Laird or me.

With an arm around each other and a can of beer in the hand that was free, the survivors didn't look as much like ghosts as they had when they went up the hill, but at water level they were still impressive and I was still full of wonder. Among other things, I kept wondering if some of the big things we had done on the hill were wrong, even if they had been done by the best four men in the business.

10

Coming to recognize you are wrong is like coming to recognize you are sick. You feel bad long before you admit you have any of the symptoms and certainly long before you are willing to take your medicine.

I felt another trip to Mann Gulch coming on. Harry Gisborne had taken only one trip into Mann Gulch and hadn't made it back. I had three trips to my credit, one each for the last three summers. If I had to make a fourth, I would be exactly twenty years older than Gisborne had been when Jansson left him lying on his back with the moon moving across the lenses of his glasses.

Although I was reluctant about coming to Missoula this particular Friday, I actually came to town earlier than usual so Laird and I could report to Edward Heilman on the condition of the crosses in Mann Gulch. When we had been in Mann Gulch on July 1, all of them, we reported, had been inspected by at least one of our party and were in acceptable condition for the time being. I expressed the hope that the crosses would be inspected in another five years, since concrete crumbles fast on a hill exposed to record extremes of weather. He assured me that such an inspection would be made, and Laird spoke up and promised he would make the first five-year inspection plus a follow-up as often as needed. On this morning when some of the key stones seemed to be giving way in the small memorial of knowledge I was hoping to erect to the dead in Mann Gulch, it was somewhat comforting to be assured that at least the crosses would not be allowed to crumble away.

When we sat down to lunch, I said, "It's not right." Laird said, "I know it, I know it." Neither of us had to ask what the "it" was.

I went on without pausing: "The base of that tall standing tree is too close to the reef—it's probably no more than sixty or seventy yards from the crevice—and that makes where Dodge started his fire too close to the top. All testimony says it was about two hundred yards. That's what the Board of Review said, and they were only quoting what the survivors said."

"Besides," Laird observed, having been a foreman of Smokejumpers himself, "it doesn't make sense. It's hard to believe as tough a foreman as Dodge would call it quits as close as sixty yards from safety."

Laird may have asked the question, but I think I asked it, because I had gone into Mann Gulch hoping to find out not only more about fire and about death by fire but also more about the life afterwards of those who almost die by fire. So I think I was the one who asked, "Do you think they could have succeeded in trying not to remember?"

I knew, of course, that Jansson had spent much of the rest of his life remembering. I had even heard that he told the Board of Review he would answer questions about his retrieval of the bodies only if he could testify with his back to them so that they could not see his face. The two survivors could be a very different case—they were young, one even underage, and they had almost died with almost all their life yet to live. Maybe when you almost die almost before you live there is a mechanism in you that makes you reduce your memories of death so most of life will not be based on death. I was making a world made up of guesses.

It was Laird who said, "It's sure strange that they remembered exactly where Hellman's cross should be from at least half a mile away."

He added, "I can tell you something else that seems strange about that dead standing tree and the hillside."

I replied, "You tell me yours and I'll tell you another and then let's both go back to work."

He said, "If Dodge lit his fire at the base of that tree, then four or five of the crew, certainly Reba and Sylvia, died long before they caught up to Dodge. Their crosses are not nearly as far upgulch as where Rumsey and Sallee say Dodge stopped and lit his fire. But Dodge thought they all passed him."

Laird looked at his watch and asked, "What were you going to tell me before I went back to the office?"

I told him, "I was going to tell you that I don't think Dodge would light his fire in a depression, and the base of the highest standing tree is in a depression between two lateral ridges."

I had a reason to back up this conviction that I thought would appeal to Laird as a foreman who had led Smoke-jumpers to fires: Dodge would not light his fire in a depression where his men could not see him. His men were behind him and were evidently just coming out of the edge of the timber when he lit his match. Sallee is undoubtedly right in insisting that the origin of Dodge's fire has to be located where from the crevice above he could see Dodge lie down in it, because he saw him. But the seeing argument works both ways—Dodge's fire also has to be located where, from behind him, Dodge's crew could see him light it.

Laird was late getting back to work. The days were turning hot and dry. Then came the lightning storms and the fires that show up by next morning. So it was a long time before he and I got together again, although I know he was happy in the meantime. He had been ejected from his desk chair in the Regional Office in Missoula to be foreman of an emergency crew on a big fire on Scapegoat Mountain.

At our next lunch together I told Laird, "I've been thinking over what you said about Reba's and Sylvia's crosses having to be upgulch from Dodge's fire, and now I know how we should go about checking on the location of the escape fire when we go back into Mann Gulch next summer."

"Fine," Laird replied, "and I've a better plan for going into the gulch than any we've used before. I think it will be easier," meaning without saying so that he thought it would be easier for me.

Actually we were starting again a long way from where we had left off. In the interval each had gone on with his thinking about our problems and was assuming that the other shared his new conclusions. We both were starting again on the assumption we were going back to Mann Gulch next summer, although we had not parted with any such agreement, and the fact is that both of us had hoped our trip into Mann Gulch in 1978 with the survivors would be our last one. For us, a trip to Mann Gulch meant about 150 miles over the Continental Divide dragging a boat with a motor that might not work. After that came the 76 percent slope having no shade but lots of rocks cracking apart with heat and rattling with snakes. Yet we began by discussing Laird's new way of going back into Mann Gulch.

"I am thinking," he said, "that this time we should come in the back way, over the head of the gulch from Willow Creek where there used to be a kind of a road. If I can get my truck nearly to the top of the divide between Willow Creek and Mann Gulch, then a lot of the climbing will have been done for us and we can sidehill from the head of the gulch down to the crosses."

"Sounds fine," I said, and talked as if we were already on our way back. And with some dehydrating alterations, this is the way we went when the time came.

While Laird had been constructing a more efficient way to reach Mann Gulch, I had been working on a fresh method to locate the origin of Dodge's fire when we got there. Until it was located, I would have no common point from which to measure times and distances converging and diverging. The origin of the escape fire was the one place on the ground that they all touched at the end.

It is possible to retrace some of the steps in thought that led us to Dodge's fire. Of all objections Laird and I had to locating the origin of Dodge's fire at the base of the highest standing tree, the one most convincing to me was that it would mean the men whose crosses are lowest on the hillside were a long way behind Dodge when they died. I didn't believe this and was soon asking myself how I could find solid evidence against it. I thought again about the location of the crosses, because the location of the crosses is as close to certainty as anything on that ambiguous mountain. The crosses were placed where the bodies were found, and Jansson himself left a pile of rocks with a note underneath identifying each body and recording how the identification was made. Probably only the location of Hellman's concrete cross is far wrong, but then Hellman died in the hospital and there was no pile of rocks with a note underneath to mark where his cross should be placed.

This line of thought soon leads to the question, Which of all these crosses is most reliably located? and the answer to this one depends entirely upon remembering the testimony about the fire. So that the answer can be visualized from the photograph of Reba's cross, the answer is Sylvia's cross, which stands next to a big flat rock about two hundred yards straight uphill from Reba's cross and slightly outside the left border of the photograph.

Certainty requires such a rock beside Sylvia's cross. Sallee remained somewhat in doubt about his own location of the origin of Dodge's fire because there is no big flat rock close enough to it to be the rock where Sylvia was found, and Sallee is the one who should best remember the rock. Exhausted as he must have been, he kept going all the night of the fire until he led the rescue team first back to Hellman and then across the divide to Sylvia tottering on that flat rock, pleading, "Don't look at my face. It's awful."

The process of thinking for some may run smoothly, but mine, insofar as it is observable to me, is more like one of those little mud geysers in Yellowstone Park, alternating between bubbles of mud and a puff of smoke (and mud again). I puffed smoke as I concentrated on Sylvia's cross, and when the smoke had cleared away, Sylvia's cross had connected in thought with what I had long been trying to find—Dodge's fire. In bubbles had come the memory of Dodge's testimony about his actions after the main fire had passed him by—he had stood up in the ashes of his own fire; had heard a voice that seemingly was a long way off but turned out to be only a short distance upgulch and below him; and following it, had found Sylvia, whose body was so badly burned Dodge had placed him on a big flat rock out of the ashes. Dodge's memory was very precise, including his estimate of the distance between where he had been lying in his own fire and where he had found Sylvia: "Upon investigation, I found Sylvia approximately 100 feet below and 150 to 200 feet to the east of my location."

So, after Laird had finished his colorful description of our coming entrance into Mann Gulch riding in a chariot with four-wheel drive, I said, "And, when we get there, let's head straight for Sylvia's cross, and be sure we have a 100-yard tape measure with us. Since we'll be starting from Sylvia's cross to find where Dodge stood up in his fire," I told Laird, "we'll be doing Dodge's estimated distances backwards—first 100 feet upslope from Sylvia's cross, then 150 to 200 feet to the left toward the mouth of the gulch."

Robinson said, "A marker might be there."

I said, "It could be, but may be no longer."

We both knew, of course, that concrete crosses had been placed to mark the dead, but we also knew that initially, a few days after the fire, temporary wooden crosses had been erected and that the origin of the escape fire had also been marked. Its location almost certainly would have been accurate. Dodge returned to the scene of the fire the afternoon after it had passed by and remained there until all the bodies were found. Part of this time he spent in showing Jansson the place where he had lit his fire, which was then marked with a pile of rocks and a wooden cross. Laird and I assumed that this wooden cross had long ago disappeared, and certainly it was nowhere to be seen when we walked along the slope with Rumsey and Sallee.

"But if the cross is still there," Laird said, "it would be sure proof."

It was a long lunch and a long time before we got back to Mann Gulch. On the way toward finding the truth there is a lot of mud in the geyser between the bubbles and the smoke.

———

It was even later than planned before we returned to Mann Gulch. We had planned to be there by the first of July before the temperature started to set records, but these plans were made before the ex-foreman of the Smokejumpers remembered this was the time of year when salmon run the rivers of British Columbia. Accordingly, he weakened his shock absorbers over two thousand miles of Canadian dirt roads only to discover that fifty-pound salmon have no trouble breaking sixty-pound test line. I found myself pleased to tell him on his return, "Hell, you're too old to believe fishing-tackle manufacturers. On the Blackfoot, where I don't expect to catch trout over two or three pounds, I always use eight-pound test leader." I added, "Even then, I stop a couple of times a day to tie a fresh leader."

So when we finally started for Mann Gulch on July 24, 1979, it was 94 degrees in Helena, not quite so hot as the record 97 on the day of the fire but hot enough to prompt us to divide jobs toward the end of the afternoon to get the day's work done. I was to sidehill to the crosses and end the day by hitting the top of the ridge and following it back to the head of the gulch, there to meet Laird, whose late afternoon mission would take him up the bottom of the gulch. On my way back I quit worrying about dying from a heart attack. Even before I reached the top of the ridge, death from dehydration seemed more immediate, but even so I knew that, thanks to the variable winds that abide on the tops of ridges, I couldn't be in such imminent danger as Laird—in the bottom of the gulch not a thing stirred but Laird. So when we met at the head of the gulch, both of us just short of death, I asked him out of curiosity, "How hot do you think it is in the bottom?" "Between 120 and 130 degrees," he answered. "Now, be serious," I said, "and remember 140 degrees is getting toward lethal." He said, "Between 120 and 130 degrees," and he has been in a lot of heat in his day. "But you remember," he added, "that temperature is taken in the shade, and there is no shade in the bottom of Mann Gulch."

In many ways the trip turned out to be all we could have hoped for, partly because the heat gave us some sense of what men suffered in Mann Gulch on August 5, 1949.

We could never have realized our plan to come into the

head of Mann Gulch by way of Willow Creek without the kindness of the Montana Fish and Game Commission. They not only gave us permission to enter by way of Willow Creek, which is on the old Jim McGregor ranch the commission bought for a game preserve, but furnished us with saddle horses and trucked the horses to the end of the road, where our four-wheel drive could go no farther. Maybe I could have made it into the gulch without a horse, but my body would have had to wait for a helicopter to bring it back.

Even on the way in, I was given a disturbing glimpse of what I was going to suffer for having done so much easy climbing with my old Dunham climbing boots—without noticing it, I had worn the lugs on their soles too smooth to climb a 76 percent slope slick with dry grass. On the way in, while spelling my horse by leading her through a rocky stretch, I slipped and fell under her legs. I could see it happening in all its parts, part by part. I could see I was falling uphill under her legs and I seemed to have time to try to reverse the direction and fall away from her, and then, fast as it was happening, I had time to see I was powerless to make any decision and time still to see I couldn't see what her back legs were doing but fortunately time to see that her front legs were curling as if her bones were soft and were curling to bring her hoofs down on the hill just above my head. After all this, there was time to start loving my horse.

This was just a premonition of what was coming in the afternoon when I had to climb the 76 percent slope, pulling myself from one fistful of grass to the next. Even so, I did not suffer as much from my hands as from thirst.

When we reached the head of Mann Gulch, we stood looking for a tree to tie our horses to, nearly all of the trees, burned thirty years before, having rotted at the roots and fallen. So we had to settle for a couple of rotten fallen trees which might possibly hold, and we tied our horses to them, but all day I watched my horse at a great distance, expecting to see her pull out for home. She was such a good horse that I am sure she could have pulled loose. But she didn't, and it must have been terrible to have been tied there in that heat.

Paul Lloyd-Davis had come with us on this trip to help run our hundred-yard tape measure. He is an ex-firefighter, an old friend of Laird's, and at the time was a reporter for a Missoula television station, and we were glad to have another ex-firefighter along to share our sufferings.

Lloyd-Davis and I started sidehilling toward the crosses from the head of the gulch with what had developed into a two-prong plan to close in on the origin of the escape fire and to leave no doubt behind. One prong was to start at the most reliable mark on the hillside, Sylvia's cross, and from it measure backward to Dodge's fire, using Dodge's estimate of the distance from where he lay down in his fire to where he found Sylvia afterwards beside his flat rock. The second prong had been added to our original plan of attack in the winter. I had been rearranging my files, as I do now and then to remind myself what is in them. In a file marked "Yonts, Susan: Washington, D.C.," I found a photograph of the site of the Mann Gulch tragedy taken on August 16, 1949, with the origin of Dodge's fire marked on it. At the time Susan Yonts found this photo for me, she was working in the Forest Service's Office of Information in Washington and I was just starting a serious study of the Mann Gulch fire. Before long, I happened to mention to her that I was the son of a Presbyterian minister and she said she was the daughter of one herself, and pretty soon she knew as much about the Mann Gulch fire as I did, maybe more. And certainly, she soon had produced more documents about the fire out of various archives than I could understand or knew what to do with. This photograph, which is reproduced on pages 12–13 of the photo gallery, I had enough instinct to take with me.

Whoever made the notations on this photograph had trouble spelling "received," and we were almost certain that the notation "Rumsey–Sallee–Hellman crossed here" does not point to the right crevice. But we were guessing that the location of Dodge's fire in the photograph might be correct because in the file with it was another photograph which, in the same handwriting as "received," records the same date, August 16, 1949, and the reason for the occasion: "Men in foreground are placing crosses." So August 16, 1949, is the day of the wooden crosses, one of which had been placed where Dodge lit his fire.

Laird immediately became attached to the photograph. He is many fine things, including a fine photographer. But first of all he is a woodsman, and you aren't a woodsman unless you have such a feeling for topography that you can look at the earth and see what it would look like without any woods or covering on it. It's something like the gift all men wish for when they are young—or old—of being able to look through a woman's clothes and see her body, possibly even a little of her character. This may help to explain why there are few really good woodsmen and many who hope to be.

Laird said, "Let me study that photograph." And so we had the second prong to our attack. While Lloyd-Davis and I were to sidehill from the head of the gulch toward Sylvia's cross with the tape measure, Laird would cut directly for the bottom of the gulch, locate the spot where the August 16, 1949, photograph had been taken, and see if from there he could match the photograph's location of the escape fire with the topography. We hoped his route would take him to where Dodge's testimony would take me.

We both knew that when and if Laird found the photograph and the topography matching, he was not going to see anything that on first glance would look much like the photograph. In the photograph many of the trees killed by the fire are still standing. By now most of the trees had rotted and fallen, and there had been almost no regrowth in the thirty years since. So Laird would be looking at a body with almost no covering, a long dead, unburied body.

It should now be possible to accompany both Laird and me on our separate ways with our different evidence guiding us, we hoped, to the same place. To accompany Laird, you should first study this photograph as if you were a forensic pathologist preparing to identify a victim in court by matching two views of the dead body. The first view (the photograph) is of the body when it was still clothed, so you are forced to look through the clothing to match the second view, the view of the skeletal remains (the ground as it is today).

As a forensic woodsman, you will have a fairly good idea after studying this photograph of what you should be looking for even before you enter the gulch. At least, you might try to test yourself as a woodsman. It becomes a case of getting from the photograph the line that you will follow to the wooden cross (you hope). What you should first look for on the ground are those three white rocks in the lower right corner of the photograph, marked "Cargo Spot"—they are on the edge of a finger gulch and form a line somewhat parallel to it. The three white rocks, unlike most of the standing trees in the photograph, should still be there and comparatively easy to see. They should be your "location point."

The three white rocks serve as a rear sight on your rifle. To find your target, you have to turn back to the photograph, run a line from the three white rocks through the cross that marks where "Dodge Set Fire," and then continue the line up the ridge until it passes close to something in the photograph that is still there on the ground and is big and highly visible.

Such a line, when extended, almost runs between two giant rocks on the reef.

Then put the photograph back in your packsack to keep it from getting soggy with perspiration and start climbing toward the two rocks. You will know you are starting right if you cross two finger gulches, one straight and fairly deep and the next a curved offbranch of the first. Shortly afterwards, if Sallee remembered correctly, there should be a tall lone tree—fallen now, but if your line has been straight you should spot it in the grass when you get close. Beyond, there should be many fallen trees, as there are many standing trees in the photograph, confirming Sallee's testimony that, just when he and Diettert and Rumsey came out of a patch of timber after dropping their tools, they looked ahead and saw Dodge bent over lighting his fire on a grassy slope.

If you now join Laird and follow such a diagonal of topographical remains from three white rocks on the lower slope of the gulch to a pair of giant rocks on the reef, I hope that on the way you will stumble upon me waiting for you near a fallen dead tree on a grassy slope. With luck, there will also be a wooden cross somewhere close by in the grass.

———

I was waiting. I could see Laird working slowly up the hill in my general direction. There were three or four obvious reasons why he was coming slowly, but only one why he never looked up at me. He felt it would be cheating to know in advance where the course he was following should lead him.

Lloyd-Davis and I were sitting where our tape measure had taken us from Sylvia's cross. We had stuck together, partly because it takes two to run a hundred-yard steel tape measure on a hill where you need at least one hand to hang on to the grass. Also, if Lloyd-Davis had gone with Laird, he would just be going through the bottom of a bottomless gulch; with me he would be walking through the length of the crosses, from the Four Horsemen who ran farthest and fastest to those who gave out soonest.

I walked through the lowest tier of crosses, heading as directly as possible to my memory of Sylvia's cross; Lloyd-Davis moved from cross to cross just above me. The crosses are almost indistinguishable from the bunch grass dried to hay and the gray-white crumbled sandstone washed down from the reef through the millennia. Sometimes you come to the cross-after-next before you realize you have missed one, and nearly always you have to part a wreath of wild hay to find the name in bronze. We shouted the names back and forth to each other, usually upper tier to lower tier and back again: David Navon called down and Marvin Sherman called back; Leonard Piper called down and Robert Bennett called back; then James Harrison called down and I knew we were almost there. The names echoing up and down the hillside might have been a roll call in Arlington National Cemetery: Navon (first lieutenant, 101st Airborne Division); Sherman (seaman second class, United States Navy); Piper (also United States Navy); Bennett (United States Army medical technician). As the hillside became ceremonial, old lines of poetry started through my head: "In Flanders Fields the poppies blow / Between the crosses, row on row. . . ."

I couldn't make myself go any farther in the poem. Nothing but the crosses fitted. "Poppies," hell. Everything was dead on the hillside—crumbling concrete crosses, rotting trees, and dead grass. As for "rows," there was a cross anywhere a body ran out of brains for lack of oxygen and rolled downhill into black death—if it was lucky, after being dead.

Already I was getting dehydrated and afraid I couldn't

think straight. However, enough oxygen somehow got to my brain to enable me to observe that we hadn't seen a rattlesnake on this snake-infested field and to conjecture that rattlesnakes, being cold-blooded and therefore sensitive to extremes of temperature, were in their holes to keep from frying. It is not possible away from that hill to feel its heat—it was hot enough so that I was becoming more and more afraid I could not think when I needed to. It frightened me that this was probably my last trip into Mann Gulch and my last chance to find out the truth of its tragedy. I kept myself going by reminding myself that the only poem I had a chance of writing about the Mann Gulch fire was the truth about it. I kept saying to myself, "Remember, you've got to keep thinking straight even if you're too dry to swallow." I added, "Or to recite poems." I added a final truism for myself, "True poems are hard to find."

Lloyd-Davis and I met at Sylvia's marker next to the big flat rock. Even after I had warned myself not to become overpowered by memory, I kept picturing Sylvia tottering on the rock and occasionally calling for help and then lapsing into unconsciousness so that it would take the rescue team a long time at night to locate his occasional conscious cries. We had only enough water left for a short drink apiece, so we decided to save it since we were sure to be in worse need later on. Instead, we started with the steel tape. I took the front end of it and went straight upslope. Lloyd-Davis stayed at Sylvia's cross to stop the tape when I had dragged out a hundred feet of it.

Since we were measuring Dodge's estimates backwards, we first went 100 feet upslope from Sylvia's cross and then 150 feet laterally toward the mouth of the gulch. I made a pile of rocks where Lloyd-Davis called out "One hundred feet," waited until he climbed to where the pile was, and then started laterally toward the mouth of the gulch as near as I could on the same contour for 150 feet, where I made another pile of rocks and sat down. Lloyd-Davis remained at the first pile of rocks so Laird could see where both of Dodge's estimates had landed us.

While Laird kept working uphill in a general direction toward us but not looking at either of us, I wondered whether climbing straight upslope or sidehilling constituted the more cruel and unusual punishment, and decided that there was no difference in the degree of punishment, only in the direction from which it came—when I was climbing upslope in my smooth, worn climbers, I fell downhill, and when I fell sidehilling, I fell slightly uphill. Already weakened from falling uphill and downhill, I wondered if I could ever get back to my horse. I was afraid by now to look and see if she was still there.

Otherwise, we just sat and waited on what must have been one of the hottest hillsides in what I remembered had been described as one of the roughest pieces of country east of the Continental Divide.

Laird kept coming in my general direction until he was twenty or thirty yards below me, still on his diagonal, probably still heading for the two giant rocks on the reef. Then he looked up and saw me. When he did, he changed his course and came straight upslope. I didn't like it that he evidently had to change his course to reach me. Finally, he sat down beside me and gulped but could not swallow or speak.

When words could get out of his throat, he said, "No, we don't quite meet." When he had recovered sufficiently to give an explanation, he said, "To have met you on the course I was following, you should have been fifteen or twenty yards farther toward the mouth of the gulch."

I told him, "I can move 50 feet more that way and still keep within Dodge's estimates. Remember, Dodge said Sylvia was about 150 to 200 feet upgulch from his fire, but on purpose I came only 150 feet so I could go 50 more feet if I needed them and still be inside his estimate."

He looked only partly relieved. He said, "Even if you were 50 feet farther downgulch, I think you still would be at least that much too high upslope to have been on the line I was following."

I wasn't quite so quick to meet that objection. In our operations, Laird had tended to designate me as the keeper and master of the documents, and I had tried not to disappoint him—at a minimum I tried to know the most crucial ones by heart. Even in the heat when everything seemed odd, I was able finally to explain the oddity of why I was sitting fifteen or twenty yards higher upslope than his angle would have projected him.

"Yes," I said, "I think we are both dead on line. The location of Dodge's fire on your photograph is almost certainly the location of the wooden cross, and it almost certainly marks where Dodge lit his fire. But I have followed an estimate that Dodge made of the distance between him and Sylvia when Dodge stood up after the main fire had passed him by. Where Dodge lit his fire and where he lay down in it are two different places and are marked that way on the 1952 contour map as two different points with two different legends—point 8 marks where Dodge set his escape fire, and point 9, farther upslope, has the legend, 'Dodge survived here.'" Laird had reservations about the 1952 map, and I had some myself. But I had come to believe in it on most important points, even if part of my belief rested on the *a priori* reasoning that it was issued at the time the Forest Service was faced with parental lawsuits charging negligence. Certainly Henry Thol, for one, would instantly have recognized any factual mistake the Forest Service made.

I told Laird, "On the 1952 contour map, the spot where

Dodge lit his fire is three contour lines downslope from where he was when he stood up after the main fire passed him by, and each contour on the map represents twenty feet of elevation."

"God," Laird said, "do you think there's a marker there?" We both looked at where a marker ought to be if it were still standing, and could see only grass. Even so, both of us would have bet that some fifty feet laterally toward the mouth of the gulch and then some sixty feet downslope was where Dodge lit his gofer match. Neither of us, though, made a dash to find out, and only slowly and almost anonymously revealed to each other what was holding us back. Eventually, but clearly, we acknowledged to each other that, if we were right in our location of the origin of Dodge's fire, Sallee could not have seen Dodge or the crew from the crevice he and Rumsey had located the summer before.

We studied the reef above us. Directly above was a wide open saddle, then upgulch a short distance from the saddle the reef appeared and had no break in it, then after the solid stretch of reef was a crevice. Without moving from where we sat, we studied the new crevice, which was on a slight angle to the right above us. Beyond it was still another wide saddle, the one which the summer before had brought us to the top of the ridge from Rescue Gulch, and after it was the crevice that Rumsey and Sallee had picked as the one through which they had reached safety.

"How far do you think it is to this new crevice almost straight above us?"

"About two hundred yards," I answered. "One hundred to two hundred yards would make it fit with the testimony in my packsack."

Laird nodded. "Maybe a little less than two hundred yards," he said. "Let's go."

We again divided missions; Laird was to go "down," and I was to go "up"—neither of us had to be more specific than that. Lloyd-Davis and the measuring tape went with me; our chief job in going "up" was to measure the distance from Dodge's fire to the new crevice. Laird sidehilled fifteen or twenty yards toward the mouth of the gulch before starting "down." "Down" was where we hoped Dodge had lit his fire—about twenty yards down.

I have had to learn a good many things to tell this story—one is how it might feel to die in the heat of the Inferno. Since the Inferno is also a pit, I have had to learn how to die in the Inferno always falling down, and always falling down I now know is a terrible way to die—it destroys the confidence before it destroys the body, and it must be terrible to die with nothing left but the body.

As Lloyd-Davis and I headed up the ridge toward the head of the gulch, I looked for Laird and saw him far off stretched out in the shade of the one standing snag on a lateral ridge toward the head of the gulch.

Even so, he got to the horses before we did, and, thinking of me, he led my horse at least a hundred yards downslope, and I held my horse by her bridle to steady myself. For a guy who should have been close to death, Laird looked pretty animated. "If you had one wish in the world," he asked, "what would you wish most?" I said without hope, "A drink of cold water." He was deeply disappointed. "Come on, now," he said. "If you had just one wish in the world, what would you wish?" Cold water still had to be the only answer, but it wasn't the right one. I just stood there leaning against my horse and looking at him. He looked back at me—pleadingly. Finally he couldn't wait any longer for me to get enough oxygen and brains to wish right. "The wooden cross," he said, trying to contain himself. "The wooden cross. It had fallen and was hidden in the grass, but it was right where it was supposed to be."

I was glad it was still his turn to speak, because I was still unsteady. He said sure enough it was a wooden cross, and he had found it almost exactly where the photograph and Dodge's testimony would have met—at the base of the largest tree on the hillside and the only one for some distance around. "So it was almost exactly where all three projections meet—the photograph, Dodge's testimony, and Sallee's memory—almost at the base of that single large tree, except it's fallen now and the grass is tall."

He added quickly, "It was lying in the grass almost hidden, but I cleaned it off and set it up and piled rocks about it. You will see when I get my film developed."

I turned my horse around so she stood below me. I needed the foot advantage that gave me to crawl into the saddle. My horse immediately started for the top of the ridge and probably for home, so I had to check her to give my speech. "It's about 175 yards from the site of Dodge's fire to the crevice," I told him. He looked pleased and said, "That's about right." I told him, "Sallee in his testimony said it was about 200 yards, and that's about what it looked like to me." He said, "It's so steep and rocky there it looks farther than it is, especially when you know you are going to have to climb it. One hundred seventy yards is probably right."

I held my horse back a moment longer. "Guess," I asked him, "what's on the other side of this crevice when you crawl through it?"

"Don't tell me," he said, his eyes wide.

"You're right," I told him. "At the far end of the crevice is a juniper bush that Rumsey could have fallen into."

We both believed in the juniper bush, and we both now wondered how it could suddenly be uncomfortably cool, as it had been on the evening of that record-hot day of August 5, 1949. We were leaving a strange world.

———

At the divide between Mann Gulch and Willow Creek I turned my reluctant horse around and took a look at Mann Gulch. It is not possible from there to see all the way to the river because of the bend near its mouth where the fire blew, jumped the gulch, and started up the canyon to meet the crew coming down. So the entity you see looking down Mann Gulch is not quite all of Mann Gulch, but the frame left of the Mann Gulch fire. It is a small, fierce, self-contained world made up of many worlds set off from the outside by fierce slopes and rocks. On either side is a different world, mountains to the south and west and plains to the north and east, different ways to live and die. The two-mile world of the Mann Gulch fire had already changed to many worlds for me, with many different feelings—the romantic earth of young men descending from the sky to find fire; the brutal earth performing a tragedy in a sentence; an earth, increasing in fierceness, refusing to yield up any more secrets of how to put things together; and always the sad world that parents cannot bear to visit. All were worlds in which I might have lived. Looking down on the worlds of the Mann Gulch fire for probably the last time, I said to myself, "Now we know, now we know." I kept repeating this line until I recognized that, in the wide world anywhere, "Now we know, now we know" is one of its most beautiful poems. For me, for this moment, anyway, my world was changed to this one-line poem. Finding it a poem, I hoped I could next complete it as a tragedy, more exactly as a story of a tragedy, more exactly still as a tragedy of this whole cockeyed world that probably always makes its own kind of sense and beauty but not always ours.

There was no water until we reached Willow Creek. I was sorry for the horses, but I was no longer sorry for us. Such can be the effect of the beauty of a very short poem.

11

Beer doesn't seem to do much to remove dehydration, but it makes it easier to admit error. On the third bottle I said to Laird, "I think I can explain how we went wrong in locating the origin of Dodge's fire last year when Sallee and Rumsey went with us into Mann Gulch."

Laird said, "I think I can too."

We were back at the Fish and Game Commission ranch, had unloaded the saddle horses from the truck, and were now leaning on the front fenders of Laird's truck, one on each fender, both of them too hot to touch but not to lean against. We were too tired to sit down in the shade, if there was any, so we put the plastic bag with the rest of the beer between us on the hot hood of the engine. We figured, since beer couldn't take away dehydration, we might as well drink it warm.

"I've thought a lot about it," I said. "I mean, I've thought a lot about it before today."

"That goes for both of us," Laird said.

For a beginning, I tried to separate out the differences in basic evidence to explain how we went wrong. Oddities explain it—oddities of terrain and oddities of psychology, although oddities of psychology aren't usually as odd as they first seem. What's really odd is how the terrain and the psychology came together in odd ways. Within fifteen minutes we had pieced together an explanation with only minor differences between us.

The first oddity that led us to error was that in 1978, for otherwise good reasons, we had approached Mann Gulch from Rescue Gulch. Approaching Mann Gulch from Rescue

Gulch is approaching it from its side, so the first thing we saw of Mann Gulch was its northern ridge. That meant, of the key points in the fire we were trying to locate, we looked first for the crevice because it was the closest to us. Psychology also guaranteed that the crevice would be our first location point—Rumsey and Sallee owed their lives to it.

But it was a dangerous procedure to use the crevice as the starting point from which to reconstruct the tragedy that happened below it. Both Laird and I had already observed an oddity at the top of Mann Gulch, a pattern of terrain that recurs along the ridge: first, there is a wide open saddle where the reef has not extruded or, more likely, has eroded away, then the reef emerges as a fairly solid cliff, then the cliff lowers and splits into a crevice or several crevices, then the reef erodes away into another wide open saddle followed again by the reef appearing as a solid cliff, which again erodes into a crevice and farther on into another wide open saddle—and so on up most of the ridge.

Given this recurring pattern of open saddle, cliff, and crevice, it was dangerous to reconstruct our map according to the first crevice we found, because there are quite a few crevices on the Mann Gulch ridge that to Rumsey and Sallee might look like the one they had squeezed through nearly thirty years ago in smoke and exhaustion.

The game trail explains at least in part why we went wrong. We were naturally following it and game naturally looks for the lowest place to cross the ridge and the lowest place nearby was the wide open saddle upgulch from us. So we reached the top of the ridge sidehilling upgulch, and upgulch is the way we kept going, although as it turned out the game trail took us past the right crevice. It was behind us when we reached the top.

Perhaps there is a greater oddity that permeated all these

little oddities, one that perhaps influenced all our movements on the field of catastrophe that day. Part of what I
wanted to find out in Mann Gulch that year was how much
two tough, able outdoorsmen would forget about the most
unforgettable experience in their lives, and I felt fairly sure
that part of the reason the two survivors came back to Mann
Gulch with two amateur experts on the fire was that they were
looking for answers to the same question. They wanted to see
if they would recognize Armageddon if it were not in flames.

Even the beginning to the answer I reached is complex
and contrary to expectation: (1) They remembered, seemingly with total recall, an incident on the periphery of their
experience in which, after they were safe, they had tried to
save one of their crew (Hellman). (2) Their errors came when
they tried to tie to reality the places most crucial to their own
salvation—that is, the origin of the escape fire where they cut
loose from their foreman and took their own lives in their
own hands, and the crevice where they saved their lives.

Almost certainly, ordinary expectations projected opposite conclusions—that they would remember most accurately the two crucial incidents in the most crucial moments
of their lives and that their memory for details would fade.
But the fox went the other way, as the fox often does when the
fox is psychology. What they remembered with remarkable
accuracy is what happened after the fire passed them by and
they crawled out of their rock slide, knowing at last that they
were safe. Their relocation of Hellman's cross by the rusted
can of Irish white potatoes is not the only example. I walked
with Rumsey part of the way out of Rescue Gulch late that
July 1 afternoon, and, with nothing we were saying to bring
this up, he said to me, "This is where I first saw the rescue
crew the night of the fire when I started for the Missouri
River to get water for Hellman."

It had been about midnight when he had met the rescue crew coming up, and the rise we were standing on was not particularly distinguishable and there are many not particularly distinguishable rises on the several finger ridges running down the slope of Rescue Gulch.

"Look down this ridge," he said, and I did. "See that big rock, with two smaller ones near it?" I could see something like some big rocks two hundred yards downridge. "Well," he said, "they were coming up this ridge. They naturally had their flashlights on, but I didn't see their flashlights because they were coming up behind the rocks. I could see the lights reflected in the sky, but the first I knew that they were behind the rocks is when the rocks appeared out of the dark and began to glow. Suddenly flashlights blinked between the big rock and the little ones, and I yelled to the rocks."

If any psychological generalization can be based on this day in Mann Gulch, it would be something like this: we don't remember as exactly the desperate moments when our lives are in the balance as we remember the moments after, when the balance has tipped in our favor and we know we are safe and have turned to helping others. Even if this is something of an explanation, it leaves unexplained what we mean by "remember." In Rumsey and Sallee's memory, the experience of their flight from death is not bound together by narrative or cartographic links—it would be hard to make a map from it and then expect to find ground to fit the map. It has the consistency more of a gigantic emotional cloud that closes things together with mist, either obliterating the rest of objective reality or moving the remaining details of reality around until, like furniture, they fit into the room of our nightmare in which only a few pieces appear where they are in reality.

There is always a good deal of housekeeping going on in our dreams—sometimes I think there is only one remembered objective detail that gives position to whatever else is "remembered" of outer reality. Most certainly our day in Mann Gulch was made to fit the crevice.

———

So it had taken us three years to locate two places on the ground—a summer to discover whether any survivors still had addresses on this earth; then a winter to induce the two still alive to return to the top of the ridge they had been trying to forget, followed by a summer when they came back and were ghosts for a day; and then still another summer for Laird and me to find out that they had been successful in forgetting certain things. Three years and two established locations; it doesn't seem like much, even though we had been following a lot of other trails during those years leading to parts of the story of the Mann Gulch fire—trails of experts on death, especially death by fire, and the many trails of Earl Cooley, who with his partner was the first ever to drop on a forest fire from a parachute and who tapped the calf of the left leg of each of those about to die in Mann Gulch to get them on with their last jump. Later I was to see the sad trail of the once majestic C-47 that had dropped its crew into Mann Gulch when the great bird of the sky circled the airstrip in Missoula and disappeared forever into the blue, sold into slavery to an African company. Stories and stories—a storyteller has all kinds of stories going at one time out of which he hopes he can find one story he can tell at one time.

I had decided that a search for the main line of the multiple story should include an attempt to determine the speed of men and of fire in their race between established points and to relate the decreasing distance between men and fire

with the tragically increasing intensity of the fire's heat. And
I did try to proceed in this quest by closely studying the survivors' testimony concerning the closing gap between them
and the fire. But testimony in such circumstances naturally
lacks exactness and even agreement, and before long Laird
and I came to the precipice of doubt with the discovery that
official testimony about the time and distance of the race
may have been tampered with. For an interval, we thought
there was no way of going any farther with this part of the
story until and unless a method could be found of ascertaining time, distance, and speed other than by pacing and grade
school arithmetic. In my often aimless visits to the Smokejumper base I almost always had to pass the Northern Forest
Fire Laboratory, where without my realizing it they were developing such a method. I had already heard—usually from
young members of the Forest Service—that in the Northern
Forest Fire Lab they made mathematical models of fire, and I
was mildly curious, although I didn't know much more about
mathematical models of fire than that they are used in the
operation of the National Fire Danger Rating System and so
have something to do with predicting the rate of spread on
any given day of forest fires real or hypothetical. Occasionally I would say to myself, "Some day when I am too old to
chase real forest fires, I will come back here and try to find
out what a mathematical model of a forest fire is, if by then
it is not too late." So I would keep on going past the Fire Lab
to where there would be a bunch of visitors watching a game
of volleyball being played with surprising ferocity by young
men who also to the surprise of the visitors were not as big as
the Minnesota Vikings. When the game was over, the young
men would return to "the Loft" and, to the ever-increasing
surprise of the visitors, would sit in front of sewing machines
and peacefully mend their parachutes. They were very skill-

ful with their sewing machines and damn well better have been, since their lives hung on their parachutes.

On a good many days in my search for the story of the Mann Gulch fire I never got any farther than "the Loft." At one end of it was a glass cage containing the Loft foreman, Hal Samsel. Hal is the son of a Forest Service ranger and was born in a ranger station and all his own professional life has been in the Forest Service, most of it in the Smokejumpers—in fact, in early August of 1949 he had just returned to the Smokejumper base from an earlier fire or his number would have been called to go to Mann Gulch. There isn't much he doesn't know about the woods, whether viewed from the ground or the sky, and he can tell stories about the woods from either perspective. He is a master storyteller, the only one I ever heard who could tell a whole story with only two grammatical subjects.

"Them sons-of-bitches," he said, opening with his first subject, "was Mennonites and wouldn't fight in the last war—said they wasn't afraid to work or die for their country but wouldn't kill anybody, so somebody, maybe for this somebody's idea of a joke, had them sent to the Smokejumpers. It turned out them sons-of-bitches was farm boys and, what's more, didn't believe in using machines no way—working was just for their hands and their horses, and them sons-of-bitches took them shovels and saws and Pulaskis and put a hump in their backs and never straightened up until morning when they had a fire-line around the whole damn fire. Them sons-of-bitches was the world's champion firefighters."

His second grammatical subject he saved for the end. "The rest of us bastards," he said, "was dead by midnight."

Many days my many-purpose quest for the Mann Gulch fire never got me farther than one of Hal's stories, but a sto-

ryteller should never look at a day as lost if he has learned
something about how to tell stories, especially about how to
make them shorter.

By the time I knew only where the crevice and the foreman's fire were in the main story of Mann Gulch, I already
knew most of another part of the story. I already knew there
would be such a part, and I knew I would search for it when
I first saw the fire itself and the black, burned ghost of a deer
bleeding where its skin had melted. I knew I would write
about this the moment I discovered we had not thrown a
rifle into the cab of the truck and so could only hope the
deer would die soon. It was at this moment that I knew my
story of the Mann Gulch fire would have a part in it asking
the question, Did any good, any good at all, come of this?

My father has a way of making his presence felt in any
story I tell, even when he isn't a character in it. He was a
Presbyterian minister and kept me out of school and taught
me himself until the juvenile officers caught up with me. In
retrospect I think the experience of listening to me recite
the Westminster Catechism influenced his own literary style,
and perhaps even mine in later times. My guess is that my
interest in this question of whether any good resulted from
the Mann Gulch fire goes all the way back to a sentence of
his, which sounds as if it comes out of the Westminster Catechism but doesn't. It is enough, though, that it sounds like
him: "One of the chief privileges of man is to speak up for
the universe."

It could be that we ask this big question almost the first
thing of all because the question seems to ask itself without
being asked. It has to be one of man's first questions after
he discovers he has personal connections with death, and I
had asked it about the Mann Gulch fire long enough to have
collected a fairly solid answer to a part of it. The *Report of*

Board of Review had divided it into two subquestions, and I had accepted the division to make answering easier. The *Report's* final recommendation was to "continue and intensify efforts in the study of fire behavior to furnish more dependable bases for anticipating blowups, and to intensify training of firefighting overhead in this respect." So I had long been asking myself (1) Did this tragic fire help to increase the *scientific knowledge* about fire behavior in ways that help modern firefighters keep out of death traps? and (2) Did this tragic fire help to improve the *training* of firefighters in ways that would add to their safety? I had the additional, private hope that something would come out of the tragedy that could circle back and help to explain what had been inexplicable about it.

The old-timer in the Forest Service I feel closest to, Bud Moore, had been top man in firefighting in Region One of the Forest Service as director of fire control and aviation management. After Bud retired from the Forest Service, he built a beautiful log cabin on the side of a Mission Glacier only thirty-five miles from my cabin, and I often go up to see him. One morning I woke with the feeling that I was about ready to write an answer to one of the first questions my father would have asked about the Mann Gulch fire. This feeling, when generalized, is a feeling that you will be ready to write if first you can find the right friend to listen to your opening paragraph. I drove the thirty-five miles to Bud's.

Before starting to talk, we took a pail down the mountainside to his spring bubbling with underground water melted from the glacier above, and I dug out a bottle of Ancient Age from the trunk of my car. When Ancient Age and ice water are sloshed around in a tin cup, the water is just as good as the whiskey.

Afterwards, I went back to my cabin to write on the effects the tragedy of the Mann Gulch fire had on the know-how of firefighters. What I wrote comes next, and Bud Moore was the first to check it.

———

The Mann Gulch fire is the Smokejumpers' lone tragedy on the fire-line. Two jumpers have since died while jumping, caught and hanged in the coils of their own jumping ropes. But the fact that the Smokejumpers have suffered no fatality from fire since Mann Gulch suggests they learned some things from it.

Those who knew something about the woods or about nature should soon have perceived an alarming gap between the almost sole purpose, clear but narrow, of the early Smokejumpers and the reality they were sure to confront, reality almost anywhere having inherent in it the principle that little things suddenly and literally can become big as hell, the ordinary can suddenly become monstrous, and the upgulch breeze suddenly can turn to murder. Since this principle comes about as close to being universal as a principle can, you might have thought someone in the early history and training of the Smokejumpers would have realized that something like the Mann Gulch fire would happen before long. But no one seems to have sensed this first principle because of a second principle inherent in the nature of man—namely, that generally a first principle can't be seen until after it has been written up as a tragedy and becomes a second principle.

In their early days, the Smokejumpers were still cautious and were still primarily limited in aim to getting on fires as soon as possible while they were small and could be put out quickly. As we have seen, the crew on the Mann Gulch fire was practically devoid of experience on big fires.

For instance, according to the *Report of Board of Review,* the second-in-command, Hellman, in 1947 had been on four Class A fires (less than one-fourth of an acre), two Class B fires (one-fourth to nine acres), and one Class C fire (ten to ninety-nine acres). In 1948, the year before Mann Gulch, he had been on two Class C fires. The almost total experience each crew member had had as a firefighter was being almost his own boss on almost his own fire where for most practical purposes he was the only one who was in a position to save his own life. One thing for sure, being almost boss of your own body and completely captain of your own soul makes you damn fast and certain of your own decisions.

Not long in coming, though, was the answer to the question, What might well happen to a bunch of early Smokejumpers when they take on a small fire that, for whatever reasons, suddenly becomes big? The answer to the question gets almost inevitable when it's asked in this form: What might well happen to a bunch of early Smokejumpers who are dropped on a good-sized fire which looks ordinary when they land but suddenly blows up? The inevitable answer has to be something like the Mann Gulch tragedy. Before long, the thing out there in nature has a way of finding the heel of Achilles.

The Mann Gulch tragedy immediately became a flaming symbol to the Smokejumpers and to firefighters generally, especially those in the Northwest. Fortunately, there are a lot of able woodsmen in the Forest Service who don't wait around for the Forest Service to do something, and it was some of these who said to me not long after the fire, "God damn it, no man of mine is ever going to die that way." Small cracks were soon filled in, especially with technical improvements. For instance, there was widespread concern about breakdowns in the communications systems that had occurred during the fire—the failure of telephone or radio calls to be completed—and much was made of the fact that the crew's radio had been shattered on the jump because its parachute had failed to open. As a result of these and similar failures, immediate and on the whole helpful changes were made, such as a simple requirement that crews must carry a backup radio. But there were deeper and more conscience-stricken improvements. Among the overhead, there was an intense heightening of the realization that at all moments on a fire their primary responsibility is the safety of their crew and that controlling the fire is only secondary. Many Smokejumper foremen have told me that since the Mann Gulch tragedy they don't make a move on a fire without first asking the question, If I go there, where can I escape with my crew if the thing blows up? And if they don't like their own answer, they don't go.

To carry out this commitment, the overhead have to do more than constantly pledge themselves to the safety of their crews. At all moments on a fire they must have a fully operational communications system to furnish them with the best information available on which to base decisions involving the safety of their men—insofar as the moment permits, there must be no failure in direct observation, scouting, or radio and telephone communications.

The training of the crews was also improved in many particulars. For instance, their physical conditioning was stiffened, and their knowledge of fire behavior, especially of large fires, was extended. Also broadened was their schooling in the differences between the behavior of fires burning in the dense forests west of the Continental Divide and the behavior of fires burning in the dry grass and shrubs east of the Divide, where little rain is left in the clouds that have been blown across mountain ranges from the Pacific Ocean.

All these things add up, but the greatest concern was to remove the contradiction between training men to act swiftly, surely, and on their own in the face of danger and, on the other hand, training men to take orders unhesitatingly when working under command. On a big fire there is no time and no tree under whose shade the boss and the crew can sit and have a Platonic dialogue about a blowup. If Socrates had been foreman on the Mann Gulch fire, he and his crew would have been cremated while they were sitting there considering it. Dialogue doesn't work well when the temperature is approaching the lethal 140 degrees.

In this delicate job of picking and training Smokejumpers so they will have almost opposite qualities, it won't do at all to pick men who accept orders without question just because they are reticent or even retarded. They have to be so smart that they know there are times when their lives depend on not asking questions. Picking and training such men is like trying to make Marines out of civilians, but the Smokejumpers have done it, and indeed the example of the Marines has helped them do it. In 1949 many of the jumpers were veterans of World War II, and twelve of those on the Mann Gulch fire had been in military service during the war. The live and the dead have joined together to make the Smokejumpers into a semi-military outfit. If a jumper now disregards the orders of his foreman on a fire, he has just made his last jump, fought his last fire, and started for camp to pick up his last paycheck.

It is worth repeating that in the nearly forty years since the Mann Gulch tragedy no Smokejumper has died on a fire-line. Some of the changes in safety procedures that helped to establish this proud record are concrete, objective safety measures, such as the addition to training courses of experience in fighting grass fires, especially on steep slopes. But the large underlying changes are more atmospheric, like being constantly aware that one risks one's life in fighting fire for a livelihood and that sometimes saving one's life depends entirely upon taking one's life in one's own hands and that at other times one's life and the lives of others must be put entirely into the hands of one boss—old lessons that throughout time have to be learned and relearned, only to be forgotten again.

The one invention that came out of Mann Gulch and was immediately made a part of the training courses for firefighters is the escape fire. It was spectacular and had saved Dodge's life and soon became a permanent part of the common knowledge of forest firefighters. One of those it has saved is Rod McIver, now the dean of Smokejumpers at the Missoula base, whose story appears in *Reader's Digest* (February 1976) under the title "Trapped in a Sea of Flame! Drama in Real Life."

In 1957, after a succession of bad fire years, Richard E. McArdle, chief forester at the time, appointed a topflight task force to "recommend further action needed in both administration and research to materially reduce the chances of men being killed by burning while fighting fires." For scientific data the task force took the sixteen tragic fires that had occurred in national forests since 1936 and looked "for threads that run through all, or most, of them." For special study, the task force selected five classic examples, one of which was the Mann Gulch fire.

The task force itself was made up of five members, outstanding representatives of the Forest Service's overhead, from regional forester, a region being the Forest Service's largest administrative unit, to ranger, who is in charge of the Forest Service's most basic unit in the field, the district. Bud Moore, then ranger of the famous Powell District on the

Lochsa River in Idaho, was selected to represent the point of view of rangers about fatal fires.

The task force developed a practical set of recommendations, and at a meeting in Washington, D.C., it was decided these orders should be modeled on the military services' General Orders. When nobody at the meeting had a copy of the General Orders, Bud Moore, who had been a Marine all over the Pacific in World War II, found a Marine standing outside at a bus stop, and together they quickly reconstructed the Marines' General Orders, which became the model for the Forest Service's ten Standard Fire Fighting Orders.

STANDARD
FIRE FIGHTING ORDERS

1. Fire weather. Keep informed of fire weather conditions and predictions.
2. Instructions. Know exactly what my instructions are and follow them at all times.
3. Right things first. Identify the key points of my assignment and take action in order of priority.
4. Escape plan. Have an escape plan in mind and direct subordinates in event of a blowup.
5. Scouting. Thoroughly scout the fire areas for which I am responsible.
6. Communication. Establish and maintain regular communication with adjoining forces, subordinates, and superior officers.
7. Alertness. Quickly recognize changed conditions and immediately revise plans to handle.
8. Lookout. Post a lookout for every possible dangerous situation.
9. Discipline. Establish and maintain control of all men under

my supervision and at all times know where they are and
what they are doing.

10. Supervision. Be sure men I commit to any fire job have clear
instructions and adequate overhead.

These orders were issued to Forest Service personnel in
the form of a training program supported by a case study
of a classic fatal fire to illustrate each of the ten orders. The
Mann Gulch fire was used to illustrate one of these orders,
but actually all the orders could have grown out of the Mann
Gulch fire except the one or two relevant only to command-
ing very large crews.

Their crosses are quiet and a long way off, and from this
remove their influence is quiet and seemingly distant. But
quietly they are present on every fire-line, even though those
whose lives they are helping to protect know only the order
and not the fatality it represents. For those who crave immortality by name, clearly this is not enough, but for many
of us it would mean a great deal to know that, by our dying,
we were often to be present in times of catastrophe helping
to save the living from our deaths.

12

After saying what I had been building up to about the influence the Mann Gulch fire had on future firefighting, I went back to work. I felt better, though, for the interlude. It is a strange thing, picking up friendships with the neglected dead, especially when you never knew any of them and also stood pledged as a writer never to sentimentalize them or pretend to imagine they were still alive. The nearest I ever came to such fantasizing was when I would imagine all the crosses on the hillside floating together and becoming one cross and the one cross becoming only a barely audible voice asking when I was ever going to get around to telling them what had happened to them. The closest I have ever come to the outer limit of friendship with the dead is when I promised myself that at the end I will use everything I know and feel to resurrect the thoughts and feelings of those about to die in a world that roared at them but obscured itself in smoke and flames.

This period in our struggling to discover the story of the Mann Gulch fire was a trying time for Laird and me. The story of the Mann Gulch fire was still mostly a concurrent set of stories, some of them leading on for a way before losing their own trails, others that had given out earlier but with luck might reveal a fresh clue with another try, and now and then a new one unexpectedly popping out of a hole almost between our feet as if we had been hunting rabbits with a ferret. "Trail jumping" would pretty much describe what Laird and I had been doing in the three years it had taken us to locate the crevice and the origin of the escape fire.

In this process, some of these separate trails little by little became longer, but strangely they also seemed to be getting a little closer together. It was something like the fire itself. The gulch at first was full of separate spot fires; it then began to fill with smoke that largely blacked out the fires and hid what was going on—as if what was coming might be a Convergence of fires Below and Above with fires Behind and Ahead. Then suddenly such a Convergence burst into view and became total in the head of the gulch; then total Conflagration rose up and swept out of the gulch. About all that has happened since is that Laird and I have occasionally returned. This is almost the way the story of the Mann Gulch fire must go if it is to follow the fire; it must all come together as it ends.

Since we always knew that the center of the tragedy of the Mann Gulch fire was a race, early in our study we had started to make an accurate map of it, much as a storyteller early tries to make an outline of his plot. This is a story in which cartography and plot are much the same thing; if the tragedy was inevitable, it was the ground that made it so. We ultimately confirmed the accuracy of the 1952 contour map in its essential details. But the cartography of the tragedy would not be complete until we studied each leg of the race and analyzed the comparative speed of men and fire in their race between established points.

Common sense should tell us that it would be a good idea as soon as possible to locate the beginning and end of the race on the map, just as it's generally a good idea for a storyteller to have a notion of the beginning and end of his plot before constructing what went on in between, although admittedly many stories seem more comfortable not knowing where they are going and never getting there. The end of the race is easy to locate on the ground—its location depends upon the hardest of evidence, the concrete crosses.

Where the race begins is a lot less certain than where it ends, because where things begin depends a great deal on prior definition and general direction. The race for the fire and the race for the men did not start at the same place, and for the first leg of the race fire and men were going in almost opposite directions. The start of the race for the fire will be taken to be point 13 on the map, where the lower end of the fire jumped the gulch to its north side and then turned upgulch to confront the crew.

In a sense, the crew began to run from the fire as early as point X on the map, when Dodge decided that the front of the fire was too hot to handle and ordered his crew to start toward the mouth of the canyon to take on the lower end of the fire, where they could jump into the nearby river if they had to. They were to proceed downgulch on contour while he and Harrison returned quickly to the cargo area for something to eat. But the real race for the men began later. From the cargo area, Dodge could see that the fire in the lower end of the gulch, instead of quieting down in the late afternoon, was boiling up. Afraid that the fire might spread and close off the mouth of the canyon, he and Harrison hastily rejoined the crew at point Y, where the foreman regrouped his men and started them on what was to be a fast trip downgulch. Accordingly, we will take Y to be the beginning of the race for the crew.

Leg 1, crew. The first leg of the crew's race is from point Y to point 6, a distance that can be known only approximately and so, for arithmetic convenience, will be spoken of as an even four hundred yards. The foreman tried to hurry his men, but it was tough going—there was no trail to follow, they were trying to sidehill on contour, and they were keeping a watch on the lower front of the fire just across the gulch. The gulch was narrow, and the fire close to them. But the crew thought of it as interesting, and Navon was taking his snapshots.

At point 6, Dodge saw ahead that the fire at the lower end had jumped the gulch (at point 13 on the map) and was already advancing upgulch toward them. He immediately reversed direction and started back upgulch. He also reversed his downgulch intention of keeping to the same contour; from here on he tried steadily to gain elevation (without cutting down drastically on speed), hoping to gain the top of the ridge and presumably the greater safety of the other side.

Leg 1, fire. The fire's first leg of the race is from point 13 to point 6, where it started to follow directly in the footsteps of the crew. The fire almost literally followed on their heels. At point 6, Dodge estimated that the fire was 150 to 200 yards away. By the time Sallee and Rumsey left the foreman behind at the escape fire and broke for the top of the ridge, Sallee estimated the fire was only 50 yards behind.

Legs 2, 3, and 4. After point 6, the race became practically identical in course and distance for both fire and men, and, although in the first leg of the race men and fire were going in almost opposite directions, we shall see that the distance covered by each in their first leg was approximately the same (400 yards) and that, as far as total distance went, the race was an even-steven affair.

The final legs of the race also appear as divisions of the tragedy from so many different perspectives that they seem to be divisions in nature. Legs in a race are at once scenes in a tragedy, each leading to a station that must be passed to reach crosses near the top of a hill.

Leg 2. At point 7, 450 yards beyond point 6, the foreman was enough alarmed by the rate at which the fire was gaining on his crew to order them to discard their packs and heavy tools. Some of them did. Some of them already had. Others would not give up their tools, and fellow crew members had to take them from their hands. When firefighters are told to throw away their tools, they don't know what they are anymore, not even what gender.

Leg 3. Only 220 yards farther at point 8 is the breaking point. Here, the foreman, having given up hope that his crew could reach the top of the ridge, lit his escape fire and tried to persuade his crew to enter it with him. Point 8 must have been each man's most important station until he reached his crevice or his cross. The men did not know it, but for most of them Dodge's escape fire was the last place where conceivably they might have been saved. Here also with the lighting of Dodge's fire began much of the dissension and legal controversy to which the Mann Gulch fire owes a goodly portion of its afterlife. Almost immediately beyond it the crosses begin.

Leg 4. The crosses spread over a wide area, and no two are close enough together to suggest that the crew itself stuck close together or even in small bunches. On the last leg of the race, from point 8 to the crosses, it was each man for himself, with no favors asked and none given, although before the race was over there must have been some asking. Of those whom Sallee, after crawling through the crevice, saw angling toward the top, Henry J. Thol, Jr., came closest to making it to the top. His cross (L) is some 390 yards from point 8. When a "representative cross" is needed, we will take that of Leonard L. Piper (G), the farthest from point 8 of the middle group—250 yards beyond the origin of the escape fire.

The total distance of the race from Y to the representative cross G is approximately 1,320 yards, or three-quarters of a mile, and tells us, as a race, that it was run at one of the most dramatic of distances, one calling for the utmost in human speed and stamina. Earlier in this century the two most honored races were the one to crown the "world's fastest human," the 100-yard dash, and the marathon, to test human stamina. Always there will be special laurels for the winners of each of these, but as the art and power of the runner have developed in modern times, middle-distance runs have changed into sprints or dashes. First, the quarter-mile run changed to the 440-yard dash, next the half-mile became the 880-yard dash, and then came the four-minute mile, and now four minutes is nothing much to brag about. Nowadays, a three-quarter-mile race tests a combination of the utmost in speed and stamina. At the end of it, no contestant trots off the field for the lockers; teammates wait at the finish line to catch collapsing runners. It is hard to imagine what the finish line of a three-quarter-mile race would be like if the last lap of it were run on a 76 percent slope of slippery grass and rock slides, on the hottest day on record.

Piper's representative cross is the closest to marking the end of this three-quarter-mile race. Henry Thol's cross, the closest to the top of the ridge, is around 150 yards farther. Of both these distances, the conventional saying must have been true—the last part was the hardest.

———

Much of the interesting business of life is learning one way or another how to represent the earth. The easiest way still to abstract short distances is by pace and (if need be) compass, but this is not as easy as it sounds and is never very accurate. It is only accurate if you have had a lot of practice in discovering what your average pace is (inches per pace) and a lot more practice in maintaining an average pace over different kinds of ground. If you are a practiced walker, then all you need is a "tally whacker" (pedometer) to count the paces, although it should go without saying that what you need most is fairly even and level ground. In Mann Gulch, especially when you have to sidehill (as you do all the way in order to follow the course of the race), it is useless—compass-and-pace is no good where you can't pace and do well just to crawl. The most accurate method, of course, is steel tape or chain, but it takes two to run it, one to hold each end, and a lot of time to operate, and unless you live close by in Helena or Wolf Creek you are going to need three days just to get part of a day in Mann Gulch. We used a 100-yard steel tape only when we measured such crucial distances as from point 8 to the crevice.

Your best friend when you feel curious about what you are walking on is usually a good map of it, if you can find one. Fortunately Laird and I had found a very good large-scale map, the 1952 contour map titled "Part of Mann Gulch Fire Area." But how do you use this map to measure the distance of the crew's race with fire? In fact you don't find it by putting your ruler on the map, ascertaining the number of inches between point Y and the representative cross, and converting the inches into miles by consulting the scale at the bottom of the map, which says that eight inches of map represents a mile of ground. When that little piece of arithmetic is completed, the figure of three-quarters of a mile, or 1,320 yards, emerges, which is the figure we have cited as the length of the race, but that's only because it was assumed most readers have learned what they know about cartography not from contour maps but from reading maps they picked up in gas stations. If you say the race was 1,320 yards, you are reading this fancy contour map as if it were a gas station map that does not represent a third dimension, elevation.

Much of the light that can be thrown on the Mann Gulch

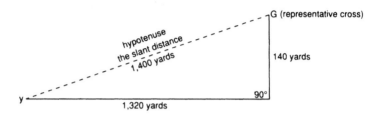

fire comes, as we shall see, from the very modern science of fire behavior, but to solve this next problem requires a classical turn back to the sixth century B.C. From point Y, where the race began, to the representative cross G, there are 20 or 21 contour lines, each representing an interval of 20 feet, and 20 feet times 21 equals 420 feet, or 140 yards. That gives some idea of how much farther the crew had to go than the 1,320 yards, but only a general idea, because the crew did not travel 1,320 straight yards on what would be a base to a right triangle and then turn at right angles and climb up 140 yards. They climbed something approximating a hypotenuse between the two points, which among map men is called the "slant distance."

The formula that gives us a slant distance of 1,400 yards at first looks incomprehensible, except that it comes out with answers that sound about right.

$$\sqrt{\left[\begin{array}{c}\text{distance on}\\\text{flat map}\\\text{between pts}\end{array}\right]^2 + \left[\begin{array}{c}\text{difference in}\\\text{elevation between}\\\text{same two pts}\end{array}\right]^2} = \begin{array}{c}\text{total}\\\text{distance}\\\text{(hypotenuse),}\\\text{the slant}\\\text{distance}\end{array}$$

If you want to go one step farther and know why it is true that 1,400 yards is 80 yards closer to the truth about the race in Mann Gulch than a flat map would suggest, you have to remember your sophomore year in high school when you were introduced to plane geometry and for the first time discovered right triangles and Pythagoras, who seemed almost to have invented right triangles. He was also very good at one-liners, and the one of his one-liners that explains the workings of this fancy formula is the one-liner he is best remembered for: the sum of the squares of the two sides of a right triangle equals the hypotenuse squared. It looks like this when A and B represent the two sides of a right triangle and C the hypotenuse:

$$A2 + B2 = C2.$$

It even works in Mann Gulch.
QED.

———

It's hard enough to find out about the things the universe prefers to keep hidden without our government, which somebody you know must have voted for, covering up what has already been found. Sometimes, of course, it hides things to save its own neck and sometimes seemingly just for the hell of it. And where does it find things to hide? Anywhere truth can be found and a dog can scratch and find something to cover it up with. Anywhere may mean as far away in the backcountry as Mann Gulch.

The timing of the Smokejumpers' race with fire was more difficult to figure than its distance, and we would need to know both before we could determine the speed of men and of fire. We had taken sixteen minutes to be the duration of the race, and at first I was fairly sure that this figure was correct. Although he said he had not consulted his watch, Dodge testified that it was "about 5:40" when he and Harrison rejoined the crew at point Y. Admittedly there is leeway in that evidence, but the time taken to be the time that the fire caught up to the crew seemed to be based on the hardest of evidence, an artifact, and an artifact is an artifact is an artifact, especially when found on the body of a dead man. The hands of the watch found on Harrison's body had melted at 5:55, 5:56, or 5:57, the fire damage making it impossible to be sure which.

The Forest Service from nearly the beginning used the melted time on this watch as marking the end of things. As the official *Report of Board of Review* says in a matter-of-fact way, "Within seconds after Dodge walked into the burned area left by the escape fire, at about 5:55 P.M., the main fire passed over. (A recovered watch stopped at 5:57 P.M.)" So in this story the in-between figure of 5:56 is usually taken to be the tragic end.

Although I was not immediately suspicious of the value of Harrison's watch as evidence, I developed an uneasy feeling when I found that, late in 1951 and early in 1952, Rumsey and Sallee had been asked to make second statements about the fire. A name began to crop up in my inquiries often enough to make me feel I had better find out who belonged to it—the second statements had been taken in the presence of an A. J. Cramer, the Forest Service "investigator" who had journeyed to Garfield, Kansas, to get Rumsey's second statement and to Lewiston, Idaho, for Sallee's.

My uneasy feeling grew when I found (not in a Forest Service archive) a letter to the regional forester at Missoula dated May 23, 1952, from J. R. Jansson. Jansson's letter is a recantation of an earlier recantation of his first testimony concerning the time of the Mann Gulch tragedy, which he refers to as the "accident." The second recantation starts by politely but painfully acknowledging that he, Jansson, had been persuaded, especially by the Forest Service investigator A. J. Cramer, to change his original timing of the race between the men and the fire so that all accounts would agree with what Jansson calls the "established time." The time of the establishment, according to Jansson, was Dodge's timing of these events. The devout Methodist then had struggled with his conscience until it forced him to recant once again and return to his original timing: "I would like," he writes to the regional forester, "to take this opportunity of enumerating my reasons for sticking with my original time statements and what supporting evidence I have to back up my conviction that my times are reasonably correct."

In addition, he implies that Dodge had been persuaded to change his original timing, which he charges was the same as his own. He says that he talked to Dodge the day after the tragedy, and, although he was too busy to "make copious notes on these conversations," he carried with him a "strong impression that there was no essential difference in our times."

Actually, the difference in time that seemed so important to Jansson may not seem like very much; as he himself says, a "time adjustment of twenty minutes in my time would bring me reasonably in line with the 'established time.'" To his thinking, "the deaths occurred . . . probably between 5:35 and 5:45." Jansson arrived at these times by reference to the times of events occurring to him at the lower end of the gulch when the fire blew up. As for the evidence on which the "established time" was based, Jansson says, "It is my honest opinion that the original investigation took Harrison's watch as prima facie evidence of establishing the time of the accident." The twenty-minute difference between Jansson and the "established time," however, would mean that the race on the hill between men and fire was over according to Jansson when it hadn't even begun according to the "established time." At 5:35 or 5:37 when it was all over by Jansson's calculations, Dodge and Harrison probably had not yet caught up with the crew at point Y.

To his charge that he was persuaded to alter his testimony, Jansson adds the equally serious charge that evidence was suppressed to give the impression that Harrison's watch was the only existing evidence to indicate when the crew was burned. But Jansson, in charge of rescue work at Mann Gulch, had "examined seven or eight watches" taken from or near dead bodies, and a year after the fire he had become deeply suspicious that this fact was being withheld from the public. Accordingly, he telephoned the regional office in Missoula (on September 27 or 28, 1950) and asked for a report on the watches in its possession. "In questioning other investigators," he informs the regional forester, "I gathered that only Harrison's watch had been found. Until the record was read to me, I had received nothing but denials as to the existence of other watches, which I knew had existed and could be read."

It was a "Mr. Kramer," keeper of the watches in Missoula, who reported by telephone to Jansson that he had four such watches "with readable times," one at 5:42, two at 5:55, and one at 6:40, a variation to think about, especially the early one at 5:42, since it can always be argued that watches with hands that stopped later were watches that kept running for some time after the fire went by.

It seems almost certain that "Mr. Kramer," keeper of the watches in Missoula, was the Mr. Cramer who over a year later was to journey to Kansas and Idaho to get second statements from Rumsey and Sallee. For the sake of everyone involved, including him, I had to find him.

I gave the documents to Laird with as few accompanying remarks as possible. I had hesitated some time before informing him about the possibility that some funny business had gone on with what he and I had come to think of as our fire. Laird was headed for a fine career in the Forest Service, and I certainly didn't want his association with me to hurt him; you've heard what Laird said to that.

The next time I met Laird, he said, "We'll have to see Mr. Cramer."

I replied, "We have to, if he's alive. He has the right of personal privilege."

"He's alive," Laird said, "or he was when I last heard. He's retired and has a home on Flathead Lake. I don't know him, but I knew one of his sons who was a Smokejumper. If the old man is like his son, he's big and tough and liable to tell you to go screw yourself."

I said, "He must be as old as I am or older, so he can't be very tough."

But Laird said, "Just the same, better let me call him. I can tell him he owes it to the Forest Service to explain these documents, and you can't make that argument."

"Okay," I said, "but first I'll write him. And I'll put it on the barrel-head to him. I'd rather be turned down in the open than sneak up on him with a smile on my face."

A few weeks later, we made a little outline of what our strategy would be for our meeting with him. Even if he agreed to see us, he probably wouldn't give us much time, so we decided to concentrate entirely on Jansson's letter and its main charges—that Jansson had been persuaded to alter his testimony concerning the time of the tragedy and that evidence had been suppressed to make a likely time seem as if it were the only possible one. I told Laird, "I have to be most interested in finding out what happened. But if it turns out the charges are true, I'd be equally interested in the follow-up question, Why to some people was it so important that the deaths of the crew appeared to happen closer to 6:00 than to 5:30 that they were willing to hide evidence suggesting any other possibility?" I asked him, "You think about it, will you? Would the Forest Service have looked like a better fire-fighting outfit if its crew had died roughly twenty minutes after it actually did?"

"No," he said, "I've already thought about it. It doesn't make any sense, even if all they were trying to do was to get all their witnesses to agree, because all their witnesses don't have to agree in a situation like this to make sense. Dodge, for instance, was at the head of the gulch closer than a hundred yards to the nearest of those who died. Jansson at the disputed time of the tragedy was probably returning from Mann Gulch in the Padbury boat. That's a lot of difference in perspective and a built-in situation to allow difference of opinion."

I told him, "Maybe our trouble is we think they thought the difference was important. Maybe we suffer from the belief that the game of cover-up is played only when there is something bad to cover up and only when big boys play it. But for a lot of guys besides Nixon it was a fun game, and all sizes, shapes, and sexes are eligible to play it.

"And don't kid yourself," I said by way of conclusion. "It is a game that can be played by woodsmen. There are plenty of big bastards who come out of the woods to become little administrators and little bastards—the woods provide no exception to original sin."

Laird always tried to look as if he hadn't heard any such remarks of mine, and I tried to look as if I hadn't made them. He said, "We've waited long enough to hear from Flathead Lake. I'll call his place this afternoon."

Late that afternoon, I dropped in to see Laird again before driving back to my cabin at Seeley Lake. Laird announced, "It's good and it's not so good. He agreed to see us, but briefly, very briefly. What I especially don't like is that he seemed old and not to understand very well what I had to say." I didn't like that either, so before I left his office we talked things over again and further shortened and simplified the questions we were going to ask him.

I met Laird at Arlee, agency of the Flathead Indian Reservation. He came up from Missoula on highway 93 and I came in a direct line from Seeley Lake on a dirt road that at its peril crosses the Mission Mountains high above the Jocko Lakes, where it almost falls into the lakes and the reflections of white glaciers still higher fall all day into the lakes. At the end of the lakes is a black canyon so buried in its steepness and shade that its gathering, unseen waters are present only as vapor rising to the tops of the cliffs where the vapor becomes visible as drops of sunshine. Then the unseen gathers noise and multiplies into a roar on its way to the Flathead Valley. Finally the roar comes into sight almost as a waterfall and, once seen, it continues as beauty.

From the mouth of the black canyon it looks as if the whole Flathead Valley had been washed out of it. It spreads like a delta of detritus, ever widening and lowering itself into golden farms.

If fertility counts as beauty, the Flathead Valley is one of the most beautiful agricultural mountain valleys in Montana, maybe anywhere. The fields, although by now all harvested, were still gold and rich enough to be used as pastureland. The cattle had been brought down from the summer grass and shade of the mountains and turned out on the harvested fields in numbers that would have overgrazed them except that the fields had been irrigated all summer from ditches starting at the edges of glaciers, so everywhere everything had followed the biblical precept to increase and multiply.

We saw little of Flathead Lake until we were almost at

Cramer's cottage, because the road on the west side of the lake only occasionally gets close to it. But what we saw wasn't much like my memories of over half a century ago when my family was thinking of building a summer home there. In Montana, they say it's the biggest freshwater lake west of the Mississippi, and big it is. But even so, it's hard to see because of the number of summer homes on its shoreline, and I was glad we hadn't built on it. We had a hard time finding the Cramer cottage among all the mailboxes lining the side road, but Laird led the way, regarding Cramer as his man. Finally he selected one red cottage out of many red cottages that look much like the mailboxes, went in, and came out with Cramer half stumbling in the lead. Age had probably shriveled him a bit, but he was still big. And it was clear he intended to keep his promise and be brief. He threw himself into a porch chair next to mine and without any introduction said to me, suspecting I was the bloodhound on the trail, "I don't know much of anything about the Mann Gulch fire. I was on another fire at the time, on a fire way up near Canada. You shouldn't bother me—you should see my oldest son, Albert. He was a Smokejumper, and he knows a lot more than I do."

Laird said, "I know your son Albert. He and I were on a couple of fires together."

Cramer was glad to look away from me, and Laird talked with him about his son and tried to loosen him up. Cramer was the remnants of a powerful man and was very tense, sure he was threatened but no longer sure by what. He was relieved to be talking to Laird about his son, but only for a moment or two, when, out of nowhere in his conversation with the former foreman of the Smokejumpers, he let me know he was thinking of me all the time by saying to me from the back of his head, "I don't know anything about the Mann Gulch fire. I was almost in Canada then." That was supposed to sound a long way off, but in northern Montana it isn't.

Having not turned around to say this, he went right on talking about Smokejumpers with Laird. I really hadn't said a word yet, not wanting to seem to crowd him, so I let him tell me twice more without looking at me that he had been almost in Canada at the time of the Mann Gulch fire before I said to the back of his head, "We didn't want to talk to you about the Mann Gulch fire; we wanted to talk to you about what happened afterwards."

He had started with the assumption that, if he could prove he had not fought the Mann Gulch fire, he had proved he could not have done anything wrong about it.

When he lost touch with this assumption, he stopped talking to Laird. "I don't know," he said to neither of us. Then he added, "I don't remember well anymore. I had an operation."

In retrospect, I would say that he was afraid of both me and himself, found both of us obscure, and hoped his wife would come soon and save him from me. Knowing I had time to ask only a few questions, I asked him, "Do you remember two survivors of the Mann Gulch fire, Rumsey and Sallee?" His fright visibly increased, but he might not have known why. He said, "No, I don't remember anybody like that." I asked him, "Do you remember the foreman on the Mann Gulch fire, Wag Dodge?" Visibly he was more frightened. "No," he said, "I don't remember well." I could not be sure his fright came from remembering these old names or from not remembering them and probably from not remembering a lot of other things. "Do you remember the ranger at Canyon Ferry, Bob Jansson, and staying at his ranger station until he agreed to change his original testimony about the time the tragedy happened at Mann Gulch?"

He looked at Laird, as if for help, and then back at me.

"Look," he said, "this has to be brief. I don't know anything about the Mann Gulch fire. I was on another fire near the Canada line at the time."

There was such a long pause that it had to be taken as an injunction to leave. Then Mrs. Cramer, who had been shopping, drove up, got out of her car, and walked around to the front of my car to see its license plate. When she saw it was an Illinois license, she hurried up the steps of the porch, introduced herself, and sat down next to Laird, probably because I looked as if I could come from Illinois and Laird didn't. She must have been deeply distressed to be caught away from home, leaving her husband unprotected from an out-of-state intruder. At first she just listened, probably trying to tell from the conversation whether we had induced her husband in her absence to say something harmful to himself. After she came to the conclusion that all the talk had been about Smokejumpers and a fire which was almost in Canada at the time of the Mann Gulch fire, she relaxed and chatted with Laird, only occasionally but skillfully protecting her husband. Her protective skill came from not making too big a thing of protecting him and from being open and matter-of-fact about things that were almost the things we wanted to know.

I heard her turn and say to Laird after she heard her husband give one of his "I-was-on-a-fire-near-Canada" speeches, "But he was an investigator of the Mann Gulch fire," and she didn't hasten to explain that admission away. She waited until another opportunity came harmlessly along to say to Laird, "He just doesn't remember very well anymore." She went on to tell Laird that her husband had had "brain trouble" and now was on medication and, as she said, didn't remember very well. Laird saw that I had heard this, as I probably was supposed to, so we would both know that the time had come for us to leave.

A flight of Canadian geese circled and lit on the water right in front of the Cramer cottage. Undoubtedly they had come from the Nine-Pipe Reservoir, a wild-bird preserve not far from the lake. The geese alternated stateliness with foolishness, then combined the two. They were stately as they circled above and as they carved the water apart for a landing and even as they stood up in their waves and slapped the water out of their wings. Then they settled back in their troughs and, just when you would expect them to reestablish the serenity of aerial motion, they broke into an anvil chorus of nonsense sounds. All of them made nonsense noise all at the same time. They headed straight for the shore and the Cramer cottage, occasionally turning to impress us with their white, majestic rear ends and then turning straight for shore again and becoming louder and more nonsensical the closer they drew. Soon the nonsense became strident and demanding and directed personally at us. I must have shown my surprise by looking at Mrs. Cramer as if for some explanation, because she said in the only remark I can remember she made to me, "He feeds the geese regularly."

"I can believe it," I replied, and soon after that we left, having overstayed our welcome by a wide margin. Even the geese seemed relieved—anyway they quieted down as we left. Cramer had gone down to the shore and was feeding them. With them and him together, the near-hysteria drained out of the scene. Soon they were talking peacefully to each other in gobbledygook.

By the time we reached Arlee the evening had drained the burnish out of the farmland but not the gold. Night was coming fast into the valley.

I said to Laird, "Thanks for the day. It's been a long time since I was in the Flathead Valley."

Like one of those youngsters who wants everything good

to happen to his older friend, Laird assumed it was his fault that we hadn't found anything, but I was the one who had started us on this mission so I had to stop him from apologizing. "What are you talking about?" I asked him. "It was a good day."

"What was good about it?" he asked, but added, "I'll never forget the geese."

"I won't either," I told him. "Scholarship doesn't always end finding a wooden cross hidden in grass."

Laird and I had prepared for the worst before we went to see Cramer. We had agreed, in fact in some detail, on what we would take to be the time of the Mann Gulch fire if Cramer was not helpful in resolving the differences about it, but it took us several days to recall the full particulars of our agreements and a lunch together to put the pieces in a new kind of order.

We had agreed first that, as far as we knew, Cramer was the last living witness who might throw new light on the credibility of the evidence regarding the time of the deaths of the crew. So if, for whatever reason, he had nothing to add, we would have to make a practical choice between the testimony of Jansson on the one hand and, on the other, the testimony of Dodge and the evidence of Harrison's watch. Such a choice has to be in favor of the evidence closest to the scene of death, not of the witness who had retreated to the river and was probably on a boat going upriver during the disputed time. Moreover, Dodge's testimony meshes much better than Jansson's with times assigned by Rumsey and Sallee to events in the gulch.

The practical choice of 5:56 as the approximate time of the death of Harrison, however, does not set aside the charge that Jansson was persuaded under pressure to alter his original testimony about the time of the climax of the tragedy. It

also does not pass over his implied charge that members of
the Regional Office had tried to create the false impression
with the public that the "established time" was based on the
only watch found on any of the dead crew. It is unthinkable
Jansson would make such a charge falsely. If he had, it is even
more unthinkable that he would have finished his life in the
Forest Service. Jansson and Dodge were both fine men and
fine woodsmen. Remember, too, that the Forest Service is a
bureaucracy, the largest in the Department of Agriculture.
That certainly makes it more than large enough for little
games.

So, actually before we started for Flathead Lake, we had
completed the practical job of filling in the old-fashioned
formula of distance divided by time equals miles per hour
to get the average speed of the Smokejumpers in their race
with fire. It was filled in as follows:

The distance of the race we had long ago determined as
accurately as we could. It was a slant distance of 1,400 yards
from Y, where the race started, to the representative cross
G, which in these calculations has been taken to mark the
tragic climax. As for the missing time, we did what you have
to do when you finally admit that you can never be sure of
the truth—you force your pride to view the spectacle of your
doing the best you can, even though that doesn't leave you
looking very good to the geese or to yourself. The start of the
race at Y, as best we know, was at 5:40. As best we know, the
end was at 5:56. So for most of the crew the race was over
in sixteen minutes, and, alas, that can be wrong by only a
few minutes.

When 1,400 yards is divided by sixteen minutes and the
result is changed from yards to miles, old-fashioned arithmetic says that the representative speed of the crew on its
journey to the crosses was three miles per hour. The *d* di-

vided by *t* and changed to miles says that Thol, whose cross is closest to the top and who traveled farthest and presumably the fastest of the crew, averaged 3.3 miles per hour—and that Sylvia, who by such calculations was among the slowest, averaged 2.5 miles per hour.

I had been somewhat surprised by these results, but Laird said he wasn't—he said they were just about what he expected. "Every guy you add to a crew," he said, "the slower you make it. After a while it scarcely moves, especially if they haven't fought fire together before." He stood there, looking back through years of smoke. Finally, in awe of the earth, he said, "A crew carrying thirty-five-pound packs in rough country will average about one mile an hour."

We talked a little about differences in the personalities of Dodge and Jansson that might account for some of the differences between them concerning the time of the catastrophe and for Jansson's charge that Dodge had changed the time seemingly agreed upon by the two of them while the fire was still burning. We were a little embarrassed by our own talk and perhaps a little embarrassed by our embarrassment. We had never known Dodge or Jansson and were certainly embarrassed to be talking behind the backs of two members of the exclusive club of fine woodsmen who could not answer for themselves.

In any event, we knew after our trip to see Mr. Cramer that we had done our best, which means that we couldn't think of anything else to do, straight ahead being a dead-end.

Trying to get another start, I circled back to general resolutions I had made to myself about getting old. I kept returning to my seventieth birthday, seventy seemingly being what man has been given as his biblical allotment on earth. I sat in my study making clear to myself, possibly even with gestures, my homespun anti-shuffleboard philosophy of what to do when I was old enough to be scripturally dead. I wanted this possible extension of life to be hard as always, but also new, something not done before, like writing stories. That would be sure to be hard, and to make stories fresh I would have to find a new way of looking at things I had known nearly all my life, such as scholarship and the woods. If you think vividly enough about your general resolutions, sometimes your conscience will furnish the particulars to exemplify them, and I became conscious again of the strange fact that on my many trips to the Smokejumper base I hadn't done much more than look inside the Northern Forest Fire Laboratory, which is next door. I had dropped in several times to see the painting of Harry Gisborne hanging in the stairwell. He always gives me the feeling that he would take a chance on trying something new, even if it didn't work, and that he is giving me the once-over to see if I feel the same way. The only other thing I knew about the Fire Lab was that a project was going on there that used mathematical models of fuels to predict the danger of wildfires and the rate of their spread. The old-timers in the Forest Service I had talked to didn't think much of this scientific project but didn't know much about it and were a little nervous, so much so they certainly weren't going to learn more than they already knew about what made them nervous. The young guys I knew in the Forest Service also didn't know much about it but thought it was great. I said to myself, "You had better be your age and learn something about it." I thought a fresh, new way of analyzing fire spread, among other things, might save me from feeding geese, and, knowing both Laird's and my mathematical deficiencies, I was sure that at least it would be hard. And it was.

13

What I remember clearly next is standing again in front of the painting of Gisborne at the Northern Forest Fire Laboratory. I have no trouble remembering back a few steps in order to explain how I got there. For a second time I had entered the lobby of the lab and again had found no one at the information desk, but by going down the equally empty hall I finally discovered an open door and was told in a somewhat dreamlike way that the two mathematicians were at a conference, in Ogden, Utah, as I remember, and the man who told me this sounded as if he were off somewhere else too. I was to discover later that being away at a conference is a basic characteristic of mathematicians wherever you find them. A lot of men have to have two places to work, one a very different place from the other. When you go into the Regional Headquarters of the Forest Service in Missoula looking for someone you need to see, you will often find that he has suddenly left to fight a forest fire in Idaho. In the Forest Service it seems as if somewhere in nature it is written, "Woe to that man who has only one place to work." In fact, that seems to be a fairly general commandment and a fairly good one to follow.

Back in the empty lobby I noticed clearly for the first time a conference table, but I didn't realize immediately that Laird and I would soon be occupying it for long periods of time. Then, recalling the painting of Gisborne, I went up the stairs to see it on the first landing, and to my surprise again Gisborne looked almost the way I thought he would, even though I was to be told later that Gisborne's eyes had been blue instead of the brown the painter chose to paint them.

But whoever painted and positioned him knew how in essence he should be represented—as an observer, and one who believed you weren't alive unless you were one too. Appropriately, he is painted full-faced, looking straight at you and looking as if looking were his business: sharp face, sharp nose, and sharp eyes (whichever color), very concentrated and aware of his powers of concentration and of how much yours could be improved.

I was glad to spend some of this day with Gisborne and the painting. Gisborne through his death kept me connected with Mann Gulch and the job I should be doing; he could connect me with the main lines of early scientific knowledge about wildfires, since he was the most important pioneer in developing the science of fire behavior. Some of these early lines of knowledge should connect ahead with knowledge about fire spread that in part resulted from the Mann Gulch fire, and some of this expanding knowledge should help to lead me to the mathematicians who I hoped would return me to Mann Gulch with a modern and exact account of the Smokejumpers' great tragedy. It would be enclosing the Mann Gulch fire in a circle of explanation.

I stood on the stair landing until I realized that the silence of the Fire Lab was being disturbed by footsteps coming down from the second floor. Actually, the footsteps and I were both glad to see each other at the landing. I was glad just then to see anybody, and, as it turned out, he was glad to see someone looking at Gisborne.

"I was a student of his," he said as he passed me.

"No?" I said in the form of a question. The "No?" in the form of a question stood for a whole bunch of things I wasn't able to utter offhand, like "You must be kidding," "Have you time for a little talk with me?" "Don't go away yet." All he said to my "No" with a question mark was "Yes."

So I met Arthur P. Brackebusch, who turned out to be not only a former student of Gisborne's but a former director of the Northern Forest Fire Laboratory. The universe only some of the time seems to be trying to prevent any discovery about it. In fact, the former director of the lab has always been gracious to me when I find him and besides I had caught him in a lull in life when he had a few moments to talk. He had been sick and had been advised to drop his full-time duties and retire to California, but not wanting to live in California he had hunted around until he found a doctor who told him California would be bad for his health. In Montana there are two kinds of doctors—those who tell you you should move to California for your health and those who tell you that you will die if you do; so Brackebusch didn't have to hunt long to get the advice he wanted. He said he would be glad to talk to me the next morning, and when I warned him that after talking to someone about a story I would need still another session with him to say back to him what I thought we both had said, he replied that would be okay, he was sure I wouldn't conflict with any medical advice he had received.

What follows, then, I said back to the former director of the Northern Forest Fire Laboratory before I wrote it, and after I wrote it he read it. I am grateful.

It was natural enough that Brackebusch and I almost immediately started talking about Gisborne—he was what we had in common. To my admiration for the way Gisborne had died, I soon added a good deal of knowledge of his life. I even knew some things about Gisborne that Brackebusch didn't know, things coming chiefly from Jansson's private insurance report on Gisborne's death, which included his last thirty-seven "rest stops." Gisborne was such a striking personality that we couldn't help being carried off into the colorful corners of his life, but Brackebusch understood, really without my telling him, that Gisborne and the science he had developed belonged in this story primarily because of the light they might shed on the Mann Gulch fire.

Although Gisborne became a model of the scientific observer and speculator at work in the woods, there was nearly always some very practical end to his scientific speculations. Ultimately, most of his projects aimed to save from fire as many board feet of lumber as possible. More specifically, most of his projects, directly or indirectly, were designed to predict the behavior of forest fires even before they got started and certainly afterwards. Even more specifically, most of these were studies to predict the rate of fire spread and its intensity measured in Btu's.

Insofar as a date can be assigned to a beginning of a science, 1922 is generally taken as the beginning of the modern scientific study of fire behavior, for it was then that Harry Gisborne was appointed forest examiner at an annual salary of $1,920 and assigned to the Priest River Experiment Station in northern Idaho as its director. The reasons leading to the Forest Service's establishment of this great experiment station clearly point out the aims and directions of early Forest Service research, directions which would never have led Gisborne to his death in Mann Gulch. The station was located on a piece of land near Priest River because all the major trees of the Northwest grew on it and in roughly the same proportions as on representative Northwest logging land, and because in addition it included burned-over areas upon which experimental trees could be planted, in other words, because it would make a fine tree nursery and because up to this time the major, almost the sole, field of governmental forestry research had been silviculture—in fact, research in the early Forest Service was located in what was called the Department of Silvics. That was the name of the game then, and perhaps oddly the fire year of all fire years, 1910, did less to promote the awaiting science of fire behavior than to intensify the traditional search for better varieties of trees to reforest a burned-over area large enough to make it eligible for statehood.

But when Gisborne came to Priest River, he was far more interested in climbing the "Weather Tree" than in planting seedlings in the nursery, an interest signifying a coming change in the direction of Forest Service research toward what was to be called the science of fire behavior. The "Weather Tree" was a 150-foot larch, limbed near the top, with platforms on the way up which could be reached by steel handspikes. As one of his scientific friends said, Gis just loved to shinny up that tree, and only doctor's orders finally stopped him, not even handspikes pulling out as the tree started to rot. On the way to the top Gisborne was interested in the behavior of the wind at different elevations and in different densities of branches. At the top were a recording anemometer, a wind vane, and a sunshine-duration transmitter, all wired to an old battery in the office, there being no electricity as yet at the station. With this primitive equipment, Gisborne was pursuing one of his major scientific aims—predicting the behavior of a forest fire. This pursuit resulted in what has become one of the most practically useful contributions to the study of fire control, the National Fire Danger Rating System, first put into operation in 1934. To keep this system in mind will help to lead us from Gisborne and the early science of fire behavior to the modern mathematicians and their fuel models and then back again to Mann Gulch.

A simplified expression of the results of this very complicated system can often be seen when one enters a National Forest. There on the roadside will be a big Forest Service sign giving the motorist the latest score on fire danger. The one on my roadside reads fire danger rating on the top line and today! on the bottom. In between goes a removable sign that a Forest Service guard is supposed to change according to fire conditions ranging from extreme, high, moderate, low, and so on, presumably to none. But near my cabin for some reason or other the fire danger never gets lower than moderate and in August gets stuck on high for days at a time, presumably because the Forest Service guard who is supposed to change it once in a while gets stuck on an emergency fire crew, and I have never seen it on low. Another ultimately simplified expression of this complicated system was used earlier in this story when it was said that on the day the Mann Gulch fire "blew," the fire danger rating in Helena was 74 out of a possible 100. That is "Danger" and lots of it.

As we shall discover when we find our mathematicians, the Fire Danger Rating System continues to attract the attention of some of the finest scientists engaged in the study of fire. The practical uses to which it can be applied are constantly extended and the accuracy and significance of its results constantly improved, since the results depend upon the close observation of complicated fire factors—temperature, fuel, humidity, grade of slope, and wind velocity—followed by the quantification of them by computers. When we talk about quantifying fire factors with a computer, we are getting closer to the mathematicians we haven't caught up with yet.

You can get some notion of the amount of scientific work needed to make the whole system function by trying to imagine the amount of study (and equipment) it took to make possible just the statement that "a fire burning on level ground (1 to 5 percent) will spread twice as fast when it reaches a 30 percent slope. The rate of speed will double again as the slope reaches 55 percent." It sounds as though somewhere around there is a computer on the side of a wind tunnel.

Gisborne seemed always to take the opposite side of whatever side you just thought he was on. He liked the spectacular and showy as well as the practical, and even wore classy leather puttees in the woods. One of his many-colored interests that brought him to Mann Gulch was lightning, and lightning was the cause of Mann Gulch's great fire. Gisborne liked lightning, as he liked crawling up tall trees to measure the wind, partly because he was a pioneer scientist and partly because he was flashy but partly because he was the son of a sawmill owner and wanted results that could be expressed in board feet. There had to be a childlike wonder in his interest in lightning and even in the difference between red and white lightning, but he was the one who more or less settled the question of whether it was red or white lightning which started forest fires; he settled it in favor of red. So even in this instance his overriding interest was knowledge that could predict the behavior of fires—in the case of lightning, careful observations which, when correlated mathematically, would let the lookouts and dispatchers know when to expect severe lightning storms and where in the mountains and at what time of day they would most likely strike.

Still another of Gisborne's major interests was the development of machinery that would either improve the observation of fire conditions or correlate observations more accurately and rapidly into predictions as to what a fire would do. He liked machinery the way a sawmill operator does, knowing his life depends upon it, and he liked it with extra tenderness because to him a machine that worked was a work of art. He became especially interested in building a tunnel for burning fires under carefully controlled influences of wind, fuel, and other factors. Important as the Priest River Experiment Station was in the development of the study of fire, it had great shortcomings that Gisborne resented for limiting his investigations. His "Weather Tree" had its points and was even something of a sporting proposition, but it offered almost no opportunities to observe fire factors under controlled conditions. It is also likely that the increasing seriousness of his heart ailment made the development of a fire tunnel even more imperative. When he received doctor's orders never to climb his "Weather Tree" again, he said, "What are you telling me? To quit?" Quit, of course, he never did. It is not even clear how closely he obeyed orders, but it is clear that before his death he had built a fire tunnel in the basement under his office at Regional Headquarters in the Federal Building in Missoula. The room had a separate chimney for dispelling smoke from his experimental fires. The great wind tunnels in the Northern Forest Fire Laboratory fulfill almost exactly Gisborne's own requirements for a tunnel. When we finally found the mathematicians at home, there were two wind tunnels in their laboratory with computers on their sides to record instantly any change in a fire's intensity and rate of spread. The whole setup is a lot fancier than Gisborne's ever was, but it is certainly something he dreamed of. And he probably dreamed in a misty way of something like the great Northern Forest Fire Laboratory, which houses as just one of its projects the mathematicians and their study of mathematical models of fuels.

The Northern Forest Fire Laboratory was built in 1960. In fact, the three fire research laboratories of the Forest Service were authorized and built at approximately the same time and, appropriately, in three of the country's greatest timber-producing regions: the South and East (in Macon, Georgia, 1959), the Northwest (in Missoula, Montana, 1960), and the Southwest (in Riverside, California, 1962). All three laboratories work on problems of national importance, but each also specializes in research problems particular to its region.

So the lab serving the South and the East originally was to concentrate on problems pertaining to hardwoods; the lab in Missoula on problems connected with lightning fires and fires in rough terrain; and the lab in California on fires that explode in chaparral and semi-arid country, or, as it was then more elegantly phrased, in a "Mediterranean environment."

The original aims of the research laboratories in California and Georgia have changed considerably through the years, but the two aims of the laboratory in Missoula remain basically the same—to study lightning fires and fires in rough terrain. These are two problems that Gisborne had made central to the study of fire behavior.

Brackebusch and other old foresters scientifically inclined believe that the death of Gisborne slowed down, at least in the short run, the advance of the science of fire behavior and the arrival of the three Forest Service research centers. In addition, the efforts of some of those in the Forest Service to bury the Mann Gulch fire in a lonely, unattended graveyard in a lonely canyon, and thus to subdue the good effects it might have had, for many years were fairly successful. Even nature seems to have supported the silence. After the burial of the Mann Gulch fire came a succession of what the Forest Service calls "good fire years," years in which there were fewer than usual fires and fewer fatal fires having to be buried. For a time the forests themselves seemingly wished that nothing more be said about sad subjects, and the universe, having frightened even itself, seemingly participated in a conspiracy to conceal its own terror.

But sure enough, then came bad years. It may not be a fixed rule but it is certainly a convention of public tragedy that it must repeat itself if it is to make a cry loud enough for something good to come of it. As for tragedy, the universe likes encores to its catastrophes and does not have to be coaxed long to repeat them. The bad fire years were a little less than a decade in coming, the climax being in 1957. To make matters worse, California was hardest hit, and when California suffers, it takes politicians to cure it.

The three labs became a story and a fact, both at the same time, because of a few politicians, two to be exact. In this respect, at least, politicians are like Smokejumpers—no fewer than two drop on a fire, and many bills are cosponsored. So it also seems in drama—two big characters are the basic minimum number. In fact, I once saw a play in which from beginning to end only two characters appeared onstage. Of course, they used the telephone a lot to talk to other characters offstage, but so do politicians.

In 1957, Richard B. Russell was senior senator from Georgia, which he had represented since 1933. Having been a member of the Senate for nearly a quarter of a century, he was one of its most influential members and a member of one of its most influential committees, the Appropriations Committee, which starts money off in the direction it is going to be spent. In fundamental ways government is as simple as determining where money is to be spent, and Georgia is a logging state with a lot of trees that can catch fire.

In 1957, Mike Mansfield had been a member of the Senate for five years after having served as a member of the House of Representatives from the 78th through the 82d Congress. Mansfield went on to become majority leader of the Senate, a position which he made into one of the most influential in the operation of our government and which he retained until his retirement from the Senate. After his retirement he was appointed ambassador to Japan by President Carter, a post he was later asked to retain by an administration bloodthirsty with Born Again Republicans. Few politicians in modern times have been more respected than the senator from Montana for their knowledge of our national government and for quiet, judicious exercise of power in influencing its direction.

In our state of Montana we would vote for him for anything (in ascending order) from dogcatcher to president of the United States to queen of the Helena Rodeo.

At this point in the story we need to take only a few steps backward in time to see the Mann Gulch connection appearing. Prior to his election to Congress, Mansfield had been a professor of history at the University of Montana, which starts to connect him both ways, to Japan and to Mann Gulch. At the University of Montana the future ambassador to Japan taught Far Eastern history. The connection with Mann Gulch starts appearing distinctly when we recall that the University of Montana is in Missoula, and thus the home town of Senator Mansfield was the headquarters of Region One of the United States Forest Service and the base of the Smokejumpers who were dropped into Mann Gulch.

The direct connection between Senator Mansfield and Mann Gulch must have been coinstantaneous with the fire. Why not? Surely the former boy who worked in the mines of Butte was at least as shaken as the rest of us by what happened to the boys working in Mann Gulch. They were two dangerous places for boys to work.

As to the Mann Gulch connection, the act had been almost as swift as the thought. The last victim to reach Helena before dying was dead by noon of August 6, 1949, and by October 14, little more than two months later, Mike Mansfield had rushed through Congress his amendment to the Federal Employees' Compensation Act doubling the amount allowed to nondependent parents of children injured or killed while working for the federal government—from a pitiful two hundred to four hundred dollars. A rider attached to this amendment made it retroactive to include the Mann Gulch dead.

The part of this story that takes us from Mann Gulch to the three forest fire research laboratories ends this way:

1. The first of the three laboratories was built in Macon, Georgia, the senior senator's home town.
2. The second laboratory was built the next year in Senator Mansfield's home town. It is only twenty yards from the base of the Smokejumpers who flew to Mann Gulch.
3. The third was built in Riverside, California, the state that had suffered most in the last bad fire year.

So the forest fire research laboratories are where the forests and the politicians were, and it would be hard to kick at that.

My meetings with Brackebusch took some steam out of my desire to find the mathematicians immediately. As I talked to him, I began to get some idea of how much more I would have to know in order to explain the Mann Gulch fire in light of the latest advances in fire science. When Laird and I were in the woods, I suppose we thought of ourselves as educated men, if there was ever an occasion to think on such a matter, but I at least knew that my education, starting with what I got from my father, had never included much math, so I began thinking that, before meeting the mathematicians, I had better retire to my cabin at Seeley Lake and do some homework. I had no trouble gathering a small pile of articles written by the two mathematicians or about them. Often it's a lot easier to find out about important persons than to find them—especially it is not hard to find out where a man came from if his last job was on a project to determine whether an atomic-powered airplane would work. Such an experiment had been conducted near Idaho Falls, Idaho, on what significantly turned out to be the Lost River. When this experiment got lost or whatever an atomic experiment does when the government doesn't want it around anymore, some high-powered young scientists were turned loose, and Jack Barrows, who had been one of Gisborne's favorite students and was now the first director of the Northern Forest Fire Laboratory, moved fast and brought up five of them, one of whom was Richard C. Rothermel, who even as a student at the University of Washington had worked in aeronautical engineering. It took some seven years, however, for Rothermel to make the switch from making models of atomic-powered planes to making models of forest fires.

It took the young scientists another six or seven years of testing and correcting their new equipment, especially the wind tunnels, before they could be confident of the results of their experiments. One of the constant and crucial questions was whether the equipment really conducted controlled experiments, whether, when it reported changed results, the changes were solely the result of the one fire factor the scientists were studying or also of other factors they had not succeeded in eliminating. Almost as difficult was the problem of getting fire to burn under controlled conditions with enough consistency and continuity to allow the results to be measured accurately. So it was 1968 before Rothermel and his group were sure enough of what they were doing to move into the field of prediction and to assume responsibility for what Barrows had long wanted, an overhaul from top to bottom of the Fire Danger Rating System.

Even then there were problems, including, of course, the problem of gaining the acceptance of the old-timers, to whom forest fires are the ultimate reality and wind tunnels and computers are gadgets. Since old-time woodsmen change even slower than equipment, it took at least another seven years and well into the 1970s before mathematical models of forest fires and their predictions became sought after by agencies and businesses that live off the woods.

Administratively, Rothermel is now leader of the Fire Behavior Project at the Intermountain Fire Sciences Laboratory (as the Northern Forest Fire Laboratory has been renamed). Scientifically, his story is close to synonymous with the introduction and development of mathematical models of fires in the Forest Service. In 1981, in Washington, D.C., he received one of the highest honors granted by the Department of Agriculture, the Superior Service Honor Award. He was cited for "outstanding creativity in developing fire behavior prediction technology and training programs, enhancing the implementation of the Forest Service's revised fire policy." The "revised fire policy" that the developing "fire behavior technology" had made possible was a change from the policy of putting out all wildfires as soon as possible (the goal expressed in the slogan "ten o'clock fires") to the increasingly prevailing theory, called "fire management," of letting a selected number of fires burn themselves out. Although this new policy remained unnerving to some woodsmen, it has proved to have much practical value, provided it is wisely used, which has come to mean depending heavily upon the Fire Danger Rating System. A rough estimate of the financial benefit that might come from fighting only some wildfires might be guessed from the fact that in the mid-1970s the Forest Service's annual expenditure had increased to three hundred million dollars and was still rising. To the economies brought about by the policy of letting some fires burn themselves out should be added a richness of ecological benefits, or the Indians long ago wouldn't have set so many prairie fires in the autumn to enrich their pastures in the spring. To the value of fire "providing a suitable habitat for wildlife or forage for livestock" can be added the controlling of insect and plant disease. The Forest Service, moreover, has not been the sole beneficiary of letting some fires burn and even setting some. State forests and private timberlands also pay close attention to the Fire Danger Rating System and, of course, so do logging companies, especially in the autumn when they have to burn their piles of slash from the summer.

Frank Albini until 1985 was a physical scientist at the laboratory, which he joined in 1962 when one of its chief problems was to gain the confidence of the loggers. In addition to being a brilliant scientist, he turned out to have a quiet, persuasive literary style that helped to make him an effective, half-concealed salesman for the extended use of mathematical models in the woods. As a scientist, he has been from his doctoral days at California Institute of Technology a maker of mathematical models, whether for Hughes Aircraft, the Institute for Defense Analysis, or General Research Corporation. As a student, he specialized in plasma physics, and, although that field has highly specified aims and subject matter, it nevertheless reflects in its ultimate goals the Greek etymology of "plasma," which has to do with form or mold. So a-modeling he has always been.

The term "making a mathematical model" shouldn't slip by us so often without being stopped to identify itself. "Making a model" for many of us suggests a bright boy using his Erector set to make a model of the Brooklyn Bridge as a structure of girders and then leaving it on the table for his mother to take down. Of course, that's not completely wrong, except that a mathematical model of a fire is a structure of knowledge and, as Albini says, a "surrogate for reality"; its girders are quantitative generalizations about things that burn in the woods and in open country, therefore quantitative generalizations about different kinds of fuel and the influences on them of such powerful environmental factors as wind, slope, temperature, and humidity.

These girders of scientific knowledge are quantitative products of controlled observations of fire experiments and actual wildfires. The challenge is to pick the right analytical generalizations about things that will burn or contribute to their burning and fit them together in such a way that they will describe a fire that is predictable in its intensity, rate of spread, flame length, and other characteristics. Quantitative models of wildfire, then, have their practical as well as their aesthetic aspects. Making mathematical models of wildfire becomes a double pleasure, and Rothermel and Albini, who derive great pleasure from building these models, were placed in charge of refining the Fire Danger Rating System—one job is part of the other, and it is a good guess that the practical and aesthetic pleasures are not separate from each other.

Albini helps us in modeling a picture of modeling by pointing out that the "origin of this kind of approach to decision making and design is the 'preliminary design' technique used in aircraft manufacture. It is no accident that aeronautical engineers are often found in model-building jobs." He then adds, "My undergraduate training was in aeronautical engineering." And, as we remember, Rothermel's was too.

Albini was once telling me about one of his projects, which had primarily to do with predicting the speed of missiles, both those missiles made by others which he had only seen and those he was recommending be built. His comment at the end was that "it's a lot easier to predict the speed of a missile than that of a wildfire." Generally, he said, "it's easier to predict the behavior of objects made by man than natural objects." Having lived long enough to absorb a considerable number of lumps and bumps from whatever hovers around outside under the name of "nature," I said to him, "That shouldn't have surprised you."

"No," he said, "it didn't. Long ago a science teacher told me, 'The universe, she is a bitch.'" Several times since, I have thought about this sentence. It's probably right.

———

When we first sat down at the long conference table in the lobby, I was puzzled about what I must have done long ago that had led me in retirement to a confrontation with two mathematicians and two wind tunnels. Some of the explanation had to be personal. As a boy I had to confront dangers of the forest that ever since have left me dreaming I am on a fire-line and the fire is about to jump the line and will if I wake up, so I try not to. My wife's ashes, scattered on the mountain she named after herself, undoubtedly direct me back to the scene of the great tragic forest fire falling in her line of vision if she can still see me. It is probably less important that when I first saw the Mann Gulch fire it was still burning through rocks like snakes on fire. In my dream from which I cannot quite awake I still sometimes see a deer with all its hair burned off except what rims its eyelids. There is no use trying to eliminate all that is personal in order to be scientific. The long conference table at which the four of us sat was big enough to take in everything and long enough to seat eighteen or twenty. Perhaps the empty spaces had been reserved for the dead Smokejumpers and Gisborne. They belonged there but were never there, but they were never far away.

Although I have had few school courses in science, I have always tried hard to be accurate with facts. In my family we were expected to be, and, in addition, I found that being accurate with facts was a kind of game and I liked to play it. Later, when I came to know some great scientists, I found that to them science was a kind of game on a grand scale. The game Laird and I hoped to play with the mathematicians was to match our analysis of the fatal race between men and fire with their mathematical study of the same race. If nothing else, the results might tell us whether the Smokejumpers had much of a chance against this fire. Critics have always talked loosely about this or that tragedy being "inevitable," but I seldom thought any of them were, and I also thought it would make a lot of difference to everyone involved if this tragedy was.

Mathematicians are very clear writers, as one should expect; their only prose weakness, also to be expected, is that they write for each other. So it turned out to be helpful to have figured out ahead of time what some of their main plays were going to be, because then it was not necessary to know, at least immediately, the meanings of all the words. As theoreticians, they start by finding it odd that, although men had been fighting fire long before they knew how to light one, they haven't formed a theory of why it spreads. I thought this odd myself and oddly applicable to me, so I made an effort to learn about some of the first mathematicians to take up this problem. Evidently, W. R. Fons in 1946, basing his work on a theory that a spreading fire is in fact a series of ignitions, was the earliest to make a mathematical model of a fire. All the other mathematical modelers of fire whom I read also started by looking for a definition of fire spread never before given, and they ended with a definition which, when reduced to its main simplistic terms, says that a spreading fire is a series of little fires. Just so we wouldn't glide by the center of this analysis with a few simplistics of mine, I asked Albini if he would write down an explanation of the process of fire spread for me and anyone who ever reads about the Mann Gulch fire:

> As the fuel burns at a point just ignited, it releases the energy that the plant has gathered from the sun and stored up as plant tissue. The tissue decomposes as it is heated by the fire (called "pyrolyzing"), releasing combustible gases that burn as a free flame. This in turn heats the remaining solid matter to drive off more combustible gas . . . . Much of the heat is carried away as hot gas, up into the smoky buoyant plume above the fire. But much, too, escapes as radiant energy from the bright flame, returning to the form in which it was released from the sun and captured by the living plant. In the form of radiation, the energy flees with the speed of light and travels in straight rays until absorbed by matter. When it is absorbed, this energy raises the temperature of the matter which has captured it. Fine, dry plant components very near the flame are thus heated very quickly to the temperature at which they must decompose, giving off combustible gases that are in turn ignited by the nearby flame. In this way fresh fuel is added to the fire, to replace that just consumed in the flame. So the fire spreads.

Therefore, when Dodge spoke of a solid "wall of flame" behind him, 250 to 300 feet deep, he was speaking figuratively as a poet, as most of us do. What was behind him were hundreds of thousands of little fires multiplying so fast that only a computer could keep up with them.

I had to walk around this explanation of the process of the spread of wildfire several times before going on, because everything ahead comes from it. The mathematical analysis of wildfire requires a structure of thought, not just some close observations of smoke and flame, and this structure is spoken of as a "philosophy" by the mathematicians. As a philosophy it has a center from which everything flows, and the center is a definition and the definition turns out to be this explanation of fire spread. If a spreading fire is a bunch of little fires becoming many more little fires, then a lot of counting has to be done to make a study of it. Think of what a lot of counting of a lot of pine needles had to be done to come to the following conclusion:

> These equations show that the rate of spread in our ponderosa pine needle fuel beds decreased by 4.23 percent for each 1-percent increase in fuel moisture. Rate of spread in white pine needle fuel beds decreased by 4.55 percent for each 1-percent increase in fuel moisture. If the effect of moisture remains linear as moisture content increases, the ponderosa pine needles would not sustain a rate of spread at a fuel moisture of 24 percent in the still air environment. Similarly, the limit for white pine needles would be 22 percent. (Richard Rothermel and Hal E. Anderson, "Fire Spread Characteristics Determined in the Laboratory," U.S. Forest Service Research Paper)

It also helps to remember what this definition looks like when it is in operation in the Fire Lab: a wind tunnel with a computer nearby. What you don't see is what is installed inside the computer—the structure of thought being outlined here, including of course as its centerpiece the quantitative definition of the spread of wildfire. "Facts" Laird and I gave the mathematicians about the Mann Gulch fire that seemed worth considering would be given to the computer, which would consider them within the structure of thought it had been given. I suppose it is something like a creamery—a lot of things are churned around in it, and some are supposed to rise to the top as butterfat.

As the structure of thought has, as its centerpiece, a definition, so the definition of the spread of wildfire has fuels as its centerpiece. In the analysis of fire spread with fuels at the center, fuels are first analyzed as particles, some of the most important factors being their size ("the ratio of surface area to volume"), heat of combustion, and ash content. Nothing is more important about the arrangement of particles than their compactness, with the limitations of being so close together as to stop a fire or so far apart as not to let it get started. Once started, however, these combustibles are "whipped," "smothered," or "kept alive" by such environmental factors as the wind velocity, the slope of the ground, and the moisture content. Just from knowing that usually the height of the forest fire season in Montana is August, when it is hot, dry, and windy, we know something of the varying influence these forces can have on any fire or sector of it.

The pioneers of fire research, including Gisborne himself, had proceeded on the assumption that the science of fire spread could be discovered by repeated trial burns using fuel samples gathered from all over the country. Two big flaws appeared in this method that kept it from producing accurate science or useful information. As discussed before, it did not produce "controlled experiments" with built-in assurances that what was meant to be tested was in fact being tested and hence could be repeated and used as data for statistical inferences. Another difficulty was that, unless you have some good idea of what you are looking for and how to find it, you can approach infinity with nothing more than a mishmash of little things you know about a lot of little things—certainly with nothing that could have been put in a hand calculator and dropped with Wag Dodge into Mann Gulch to avert his tragedy.

It is at this point that "fuel models" become necessary

and make their appearance. They are fashioned roughly like ready-made suits of clothes. It's a case of picking the fuel model you think will come closest to fitting, having a few adjustments made in the sleeves and shoulders and always in the legs, putting the model and the adjustments in the calculator, and watching the predictions come out almost instantly, usually with a margin for error. But mathematical tailors improve with practice, this experiment the four of us were conducting being something of a test as to how much.

There were twenty-six years between the first scientific concept of the spread of wildfire and its general availability as a field of thought with practical applications (the result in 1972 of Rothermel's *Mathematical Model for Predicting Spread and Intensity of Fire in Wildland Fuels*). Seven years later Robert Burgan of the Fire Lab developed a hand-held calculator program that could predict fire spread in the field. When thought has moved from general concept and definition to practical application in the field, thought has taken shape and become a field of thought.

At the time of our talks around the conference table in the lobby, thirteen models of different forest fuels were ready for use in the field: three for grasses, four for shrub, three for timber, and three for slash. As for environmental factors influencing burning fuels, coefficients have been developed to measure their effects on fire behavior. A coefficient shows the relationship between two factors. Here, for example, is one that Rothermel had already established and that will certainly be transferred back to Mann Gulch for a scientific as well as a simple human understanding of its tragedy: "The percentage increase in the spread rate [one factor of a fire] varies in proportion to the square of the percent slope [an environmental factor influencing the fire]." This is a tragic statement; it was very steep where they died.

Near here a change started occurring in our procedure around the long conference table. The grammatical structure of our sentences changed places across the table. It had started with flurries of interrogative sentences ending with rising inflections from Laird's and my side, which were followed in a much lower key by a patient monotone of declarative sentences from the mathematicians, who knew they would have to repeat and explain their answers. The first movement of this grammatical structure was expressive of Laird's and my attempts to get into our heads, however fleetingly, what had been solidly installed in the computer, a "philosophy of wildfire." But as the grammatical structure veered around, Rothermel and Albini were asking the questions, still patiently, and Laird and I were giving the answers, still flutteringly with a hurried exchange of nondescript prose before one of us uttered the agreed-upon answer. Much of this would be put back in the computer to be combed over by what the computer knew about fire spread. I believe the trade name for this is "input," a name I suppose one has to accept for an after-effect of fire. But it also is to be taken as a sign that we were nearing the final step back to Mann Gulch—getting its exact "facts" in order to pick the right fuel models and sets of environmental factors and then make the adjustments needed for a "close fit."

Many times just the practical problems of gathering accurate data on local fire conditions can be as difficult and complex as the mathematics that follows. In this quest to seek for scientific as well as human causes of a tragedy, the speed of the wind was very important, so it is not a diversion to consider here the kind of information we tried to dig up and relate there at the conference table in order to deter-

mine the probable wind velocity (or velocities) that powered
the fire on its tragic course. The data we needed came from
three different sources: official weather reports, testimony
of those present at the fire or closely connected with it, and
our general knowledge about the behavior of winds on large
fires, especially on fires that are blowups. We knew from the
dispatcher at the Missoula base and from survivors that air
conditions at the time the plane left Missoula were turbulent
and that the plane ride to the gulch was so rough that one
jumper became too sick to jump and the others felt ill; that
the wind velocity at the gulch was so great that the plane
dropped its cargo from an unusually high altitude to avoid
entering the narrow, windswept canyon; and that the cargo
when dropped was scattered over an unusually large area
and took a fatally long time to collect. We knew the official
wind velocity at Helena at the approximate time of the tragedy (at six o'clock in the evening it was nineteen miles per
hour), but Helena is more than twenty-five miles away, on
the side of a wide valley, whereas Mann Gulch is a narrow
notch that connects with the twisted canyon of cliffs of the
Missouri where winds are compressed and are often sub-
stantially higher. From our common experience with forest
fires, we knew long before Mann Gulch that big fires add to
their own wind velocity by the whirling motion set up when
cooler air rushes into the lower part of the fire to replace the
hotter, lighter air that has risen and escaped. We knew from
several sources that this particular fire was a blowup, with
fire whirls throwing burning cones and branches across the
gulch and starting spot fires that soon were racing upgulch
toward the crew.

All the survivors testify to the great heat of the fire and
its high wind, but none left precise estimates except Jansson. "The final settling of the wind . . . was to a strong twenty

to forty mile per hour wind," and we should know enough about Jansson and the language of the woods to know that Jansson is not admitting he could not tell a twenty mile per hour wind from one blowing forty miles per hour. He is saying as precisely as a trained and involved observer could say that air conditions in Mann Gulch late in the afternoon of August 5, 1949, were highly turbulent and that the wind velocity *fluctuated* from twenty to forty miles per hour. In this story, these figures will be taken as the most precise we could find in historical documents, but when a numerical velocity of the wind at the time of the tragic race has to be made a part of the arithmetic, the figure of thirty miles per hour will be used with the understanding that it surely varied, as Jansson said it did.

With their questions, the mathematicians were gathering data about fire conditions in Mann Gulch so detailed that they could construct fire-condition maps of the critical stretches of ground in the race between fire and crew. Each map showed the critical changes in fire factors that the crew had to face as they raced from one piece of ground to another. As it happens, these stretches of ground are much the same divisions or scenes that a storyteller would mark off to show the progress of his tragic plot.

The end of the summer was coming and so was the end of the fire season, which put a temporary end to our conferences in the laboratory. But there was still time after the other three returned from the fires to get in some work before I left Montana for the year; although we were separated, the work went on and was completed in first form that winter. Laird was in Missoula and continued the conferences, and I could supply my part of the answers by letter or telephone. I couldn't pretend to keep up with Laird when we were in the field, so we had tended to specialize—he swarmed over

Mann Gulch until he could remember where even the individual trees lay rotting; I tried to compensate by carrying a packsack of historical records and by remembering most of the rest of them. I don't think we made many mistakes, although it is hard for me, even in old age, maybe especially in old age, to admit how much of the truth can escape.

14

I had first heard of this business of fighting fire with mathematics instead of Pulaskis from an old-time firefighter. He didn't think much of this as business and certainly didn't know much about it, but he said it was coming and was coming from a big building just next door to the Smokejumper base in Missoula, and he said computers did most of the work, which was counting. He said they counted burning sticks and some of the sticks were homemade in the lab. He obviously saw no great future for this business. But oddly he thought I should go out and take a look at it, and I did, because he was one of the best old-time firefighters in the Forest Service and nearly always what he said should be done I did if it had anything to do with forest fires.

Even before I went out to the Fire Lab I had an idea or what eventually turned out to be an idea. It was in the form of a somewhat smoke-obscured question: If mathematics can be used to predict the intensity and rate of spread of wildfires of the future (either hypothetical fires or fires actually burning but whose outcome is not yet known), why can't the direction of the analysis be reversed in order to reconstruct the characteristics of important fires of the past? Or why can't the direction be reversed from prophecy to history? The one great tragedy suffered by the Smokejumpers was fading out of memory before its outline had been cleared of the smoke of controversy, before the missing parts, perhaps some self-cultivated, had been recovered, before its deferred trial had taken place in public court, and before its suffering had finally been placed within the reach of the public that would like to remember and honor it with sympathetic understanding. I can't say that the idea caught on like a crown fire. More like a spot fire, it started something that smoldered and kept growing, not of course without changing direction several times.

It is fundamental for an analysis of a past fire that it should be attacked with a method that combines two methods, the predictive and the reconstructive. Prediction by its nature depends largely upon the scientific method, which, quoting Rothermel, "usually conjures an expectation for an answer to a problem that has somehow been arrived at by logical deduction. In the case of fire behavior, the logic is supplied by mathematical models." When you change from prediction to reconstruction, you combine the thirteen fuel models with every bit of information taken from observation, measurement, and the historical record. I don't suppose we were the first to make a serious effort to use this two-in-one method, the predictive and the reconstructive combined, in trying to state as accurately as possible what happened in the critical stages of a complicated and tragic fire, but as a team we were fortunate in the diversity of our specializations, and no combination of investigators would again have access to the only two survivors of the fire, whom Laird and I were able to persuade to return from far away to spend a day with us in Mann Gulch. Now there is only Sallee, and I doubt that he will ever go back again. Even so, it took us some years to analyze the tragedy of the Mann Gulch fire by both predicting and reconstructing, because only recently has the science of fire behavior developed to a level where it is possible to analyze a legendary fire with any accuracy. There must have been a good many reasons drawing me back to the job, and certainly one of them was that, the more accurately the race between fire and crew was analyzed, the more it took on the form of the plot of a tragedy emerging from concealed to complete inevitability.

The tragic convergence of fire and men in Mann Gulch offers itself as a tragic model for a graph, the modern scientist's favorite means of depicting what he wishes to present as clearly as possible. Drawn along axes of time and distance is one line depicting the course of the fire and one depicting the course of the men, and where there is a convergence of the two, graphically speaking, is the tragic conclusion of the Mann Gulch story; the two lines converging to this conclusion constitute the plot. Along each line are numbers which are turning points in the race between men and fire, and if the lines are viewed as a race the numbers mark off legs of the race, if they also have religious significance they are stations of the cross, and if they have literary significance they mark off acts of drama. If it is drama it has the same old five acts as traditional drama, but the acts are much shorter, possibly because modern wildfire allows no time for soliloquies.

The legs into which the race seems always to be divided seem always to be the same, as if nature had left natural markers there. The starting legs for the fire and the crew are different, at different ends of the canyon and going in opposite directions. The remaining three legs, however, are almost an overlap of each other, the fire now being behind the men and getting closer.

The coming account of the tragic race from start to finish is based on an unpublished paper by Richard Rothermel, which in turn is based on thirty-two pages of mathematical worksheets. By now you should know enough about how these mathematical woodsmen do business to be able to predict how these sheets are divided. The first big division is made up of four sections, a page or so on each of the most significant fire factors determining the spread and intensity

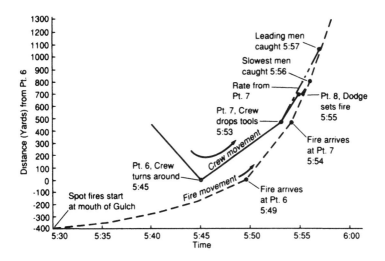

Time and position of fire and crew at Mann Gulch on August 5, 1949.
Graph by Richard C. Rothermel.

of the Mann Gulch fire: "Fuels," "Wind," "Slope," and "Fuel Moisture." "Fuels" because fuel is the centerpiece of any fire and surely was of this one; "Wind" because a roaring wind virtually assures spot fires and a blowup; "Fuel Moisture" because the fire's tragic intensity was made possible by the superdry fuels and air that came with a long dry period and record heat; "Slope" because the abrupt increase in the grade of the slope near the top of the ridge was one of the reasons why the crew did not reach it.

The second big division of the thirty-two sheets is entitled "Sequence of Main Fire," and it is here near the end that a scientific analysis of fire and a literary analysis of it as tragedy come closest to being one. In this section, each of the five legs has at least a half-page commentary pointing out the major change or changes occurring in that part of the race, each commentary being followed by a full-page "Fire

Behavior Worksheet" with twenty-three entries available to characterize the changing conditions on that leg.

It is not necessary, fortunately, to decipher all twenty-three entries to realize that, scene by scene, tragedy is approaching. Instead, let us take the word of our experts, who say fuels are the center of the tragedy; if they are, we should be able to perceive a tragic continuity joining scene to scene and increasing in intensity as we follow the changes in fuels that the crew encountered as they attempted to escape.

1. The start of the race for the fire (from point 13 to point 6). The fire had jumped the gulch near the bottom, which was narrow and "overgrown by shrubs and conifer reproduction." The two fuel models that most closely fit the conditions of the bottom of the gulch are model 10 (timber litter and understory, etc.) and model 12 (mixed conifers, slash somewhat compact). These heavy fuels under the dry, windy conditions of Mann Gulch would have created a fire of very high intensity "on the order of four hundred to a thousand Btu's per foot per second" with flames in the surface fuels from seven to ten feet high. Such conditions would generally have produced a very intense, although not fast-spreading, fire, but the strong surface wind at the time "would certainly have caused crowning and spotting, and crown fires with strong winds usually spread faster than surface fires in timber litter." The main direction of the fire would have also threatened the crew coming downgulch. Not only would the wind take it upgulch but the slope would take it uphill, so on this diagonal sidehill angle it was on its way to head off the crew, which had not seen it yet. What the crew actually saw as it approached point 6 was black smoke boiling over a lateral ridge below and ahead of them. Black, boiling smoke is often what you see before you see the crown fire that is making it.

The distance from point 13 on the map, where the fire jumped the gulch, to point 6, the turnaround, can only be estimated; we have already once rounded it off to 400 yards. The location of point 13 involves some guesswork if for no other reason than that no one ever knew how many spot fires jumped the gulch and took off. But the spot fires' location is not as important at this point as Dodge's estimate that the fire was still 150 to 200 yards away from the crew when he ordered his men to reverse course. Using fuel models and the fact that the fire had crowned, Rothermel estimates the fire was traveling at 120 feet per minute toward point 6. If we assume the fire was traveling 120 feet per minute and was still 150 to 200 yards behind the crew when it passed point 6, the fire would have reached point 6 in four to five minutes. That can be thought of as what the crew had in the way of a head start.

2. *The start of the race for the crew (from point Y to point 6).* The location of point Y can be only guessed at; 400 yards upgulch from point 6 is close and keeps the arithmetic fairly easy. The foreman was somewhat concerned, as he should have been, but it was open hillside and the men were moving gently on contour so they could keep an eye on the flank of the fire on the other side of the gulch. It would be fair to assume that they averaged about three miles per hour between points Y and 6, where they first saw that the fire was on both sides of the gulch. From point 6 on, time became a matter of minutes; 5:45 would be a good estimate of the time they reached the turnaround at point 6. There is hard evidence that at least some of the men were dying at 5:56, and the closer the race became the more accurately both time and distance can be estimated. In something close to eleven minutes the race was over.

3. *The turnaround (from point 6 to point 7).* This leg of

the tragic race, with the fire closing in from behind, was the slowest. There is always time lost just in the mechanics of turning a crew around and getting it started in another direction. But the greatest loss was the loss that came in morale and organization in turning a crew around and retreating from the fire. The training schedule of Smokejumpers includes no class on how to run from a fire as fast as possible.

The fire was having no organizational problems. It was gaining speed all the time. As it came upslope out of the brush and reproduction at the bottom of the gulch, conditions changed, and to analyze them the fuel models must also be changed. Moving out of the bottom of the gulch, the fire came into more open country with occasional large standing trees, either alone or standing in groups. The grass, according to the survivors, was at least two and a half feet tall, by August cured like hay that would cause flames high enough to torch single trees and start crown fires in standing bunches. Using model 3 (cured tall grass), Rothermel calculates that the fire had now accelerated to 280 feet per minute and that its intensity had increased to 2,500 to 4,000 Btu's per foot per second with flames from the surface fire sixteen to twenty feet high.

The slant distance between points 6 and 7 is about 470 yards, and Dodge says that at point 7 the fire was only 75 to 100 yards behind them. Rothermel's estimate of the vital statistics at point 7 follows: "If the fire were traveling at 280 feet per minute, it would cover the 100 yards in about one minute, so the crew would have reached point 7 one minute ahead of the fire, or at 5:53. The elapsed time between point 6 and point 7 was between 5:45 and 5:53 or eight minutes, a pace of 176 feet per minute or 2 miles per hour." That's the slowest time the crew made in any of the legs of its race with fire. But it is not surprising when one considers the leakage of precision that occurs when a crew has to be turned around and sent into retreat.

While the crew between points 6 and 7 averaged only 2 miles per hour, the fire already was getting ominously closer. By the foreman's own estimate, in this quarter of a mile the fire had cut the distance between them in half. No wonder that at point 7 he ordered the crew to drop its heavy tools, and when a firefighter is told to drop his firefighting tools he is told to forget he is a firefighter and run for his life.

4. Foreman lights fire ahead (from point 7 to point 8). The leg of the race from where the foreman ordered the men to drop their heaviest tools to where he lit his escape fire is barely more than 220 yards when figured as slant distance. It is the shortest distance between the stations of the cross, but it is the most critical in determining that the race would end in tragedy. This is the scene that makes the coming catastrophe inevitable, though it was not viewed that way by the tragic victims and not for some time by those who later studied it.

As the men emerged from some standing timber to see their boss light a fire ahead of them, they were also looking at a broad expanse of something but did not see its significance—the fuel had changed and greatly increased the odds against them. Ahead to where Dodge was lighting his fire and even beyond and above to the top of the ridge was a grassy clearing with only an occasional tree. It was the end of the earth for most of them who looked at it. They had come out of scattered timber with tall grass understory to an open slope covered with the flashiest of mountain fuels, cheat grass and fescue, a mixture of short and tall grass and no timber (models 1 and 3). The upgulch wind was probably about what Jansson a little earlier had estimated it to be—around thirty miles per hour—but the standing timber had kept it from acting with full force on the fire. Now it worked on the fire with full force in the open. As a result, flames in the clearing would have been thirty feet tall. No wonder that Dodge spoke of what he saw only fifty yards behind them as a "wall of flame."

On the graph the effect of the surface winds on light fuels is illustrated by the increasing steepness of the line representing the course of the fire. From points 7 to 8 Rothermel estimates the fire was traveling 360 to 610 feet per minute with an intensity of 5,500 to 9,000 Btu's per foot per second. At this speed it could cover the 220 yards in one to two minutes and arrive at point 8 at 5:55 or 5:56.

The men covered the 220 yards between points 7 and 8 in two minutes, a figure meaning that they had almost doubled their average speed between points 6 and 7. Four miles an hour was a terrific speed for the crew to maintain on that slope. At the turnaround Dodge had estimated the fire to be between 150 and 200 yards behind the crew; at point 7 only 75 to 100 yards behind. Now at point 8, only 220 yards farther, the fire had closed the gap to 50 yards. Rothermel's added testimony is that, beyond point 8, the fire had to be going about five times as fast as at point 6—660 feet per minute, or 7.5 miles per hour.

At point 8 too many questions remain to stop now and try to get answers to all of them. The central and most bitter of these questions is, Was the fire that cremated the Smokejumpers the main fire or did the foreman's escape fire keep its advanced position and reach the Smokejumpers first? Let us first accompany those yet to die as far as we can before making the escape fire stand trial, as undoubtedly it should, before it finds its resting place in history.

5. *From the escape fire to the farthest cross.* As we measure the remaining distances and times, we can divide our problem, like the grave markers, into two discernible tiers—

the front-runners, led by the Four Horsemen, who were within a minute of reaching the top, and those on the lower tier, who, after passing Dodge at his escape fire, seemingly just went on sidehilling but without the strength to climb any longer. There were no fundamental differences, however, in the fire conditions the two tiers faced.

The official transcript describes the fuels in this area around point 8 as chiefly cheat grass and fescue. Rothermel says:

I assume it to be equally divided between fuel models 1 and 3. Model 1 is the light fuel model that describes cheat grass very well; model 3 is a heavier grass model that should account for the fescue, weeds, and brush in the area. The fuel moisture calculates to be very close to 3 percent, a very low value. Probably the worst problem was the very strong wind that was gusting to forty miles per hour in the open. It was probably scouring close to the ground, giving midflame winds of fifteen to twenty miles per hour.

Rothermel's description of the fire conditions continues until it leads to his projection of how soon the men on the lower tier, and presumably the slowest among the front-runners, died:

With these conditions, and estimating the fuel was divided equally between fuel models 1 and 3, the fire is estimated to have been spreading at 660 feet per minute with flames ranging from ten to forty feet in length. At this rate, I estimate the slowest men were caught in forty-five seconds, at 5:56, just a hundred yards from Dodge. The rate of these men over this last hundred yards figures to have been near four hundred feet per minute, or about four and a half miles per hour.

That is a remarkable speed for the slowest of the fastest to maintain, especially if they lost some time at the escape fire trying to understand what Dodge was saying to them. One other piece of on-the-spot evidence was used in arriving at this calculation of their speed—Dodge's statement to the Board of Review that it would take the fire no more than thirty seconds or one hundred yards to catch the tail end of the crew sidehilling past him.

Those at the head of the line who came within a minute of the top of the ridge come close to taking us to the edge of reality. We have referred to them as the Four Horsemen; let them remain as such to the end. The average distance of their crosses from Dodge's escape fire is 375 yards. Rothermel theorizes that for them to get this far they were not with Dodge when he started his fire "but rather began their strong push at point 7." There is firsthand evidence that, for one of them, this was the case. Sallee says in his second official statement, made on December 12, 1951, that "I had noticed Navon cut off up the hill away from the crew just before we broke out of the timber and hadn't noticed him rejoin the crew." No doubt, for we have seen before that when the former parachute jumper from Bastogne didn't like the way things were going in the United States Forest Service he took matters into his own hands. But I can find no mention among the statements made about the fire by any of its survivors of the individual conduct of the other three front-runners, and I am inclined to think that the other front-runners stayed with the bunch but outran them and that one of them, Henry Thol, Jr., outran Navon.

If Rothermel is right that the front-runners began their push at point 7, they traveled about 600 yards to the site of their grave markers. He projects that their rate of travel from point 7 would average 460 feet per minute, or five and a quarter miles per hour. However, if the possibility I am inclined to believe is correct, if the other three left Dodge at point 8, and if we allow them fifteen seconds or so to watch him start his fire, then they would have traveled 375 yards in about two minutes before the fire caught up to them. "That is 562 feet per minute, or six and a half miles per hour." As Rothermel goes on to say, "That is a slow jogging pace" and would have been almost beyond reality to maintain for 375 yards on a slope where I had to crawl with gloved hands on a hot August afternoon.

All of us have the privilege to choose what we wish to visualize as the edge of reality. Either tier of crosses allows us to picture the dead as dying with their boots on. On some of the bodies all but the boots were burned off. If you have lived a life that has thrown you in contact many times with nature, you have already discovered that sometimes you can deal with nature only by allowing it to push back what until now you and others thought were its edges.

―――

No one who survived saw what happened to those who became crosses on that hillside. Until the fire had passed, Dodge lay facedown in the ashes of his fire gasping into his wet handkerchief. Rumsey and Sallee were later to testify that, as they escaped up the side of the hill after leaving Dodge, there were long flames and engulfing smoke most of the way but that two or three times the smoke lifted for a moment. Something like this is probably how it will be with us to the end of our Mann Gulch story. Outcroppings of reality will come to us only in glimpses, as they came to Rumsey and Sallee, but I know what all along I have been waiting to see. In the Forest Service reports issued after the fire there is scarcely a see-you-later-alligator farewell for the thirteen men who were killed. When Dodge was asked at the Board of Review if any of those about to die looked his way as they went by, the foreman replied, "They didn't seem to pay any attention. That is the part I didn't understand. They seemed to have something on their minds—all headed in one direction." The last Sallee could recall seeing them, "They were angling up the slope in the unburned grass and fairly close to the edge of the fire Dodge had set." Then the smoke and the Great Ambiguity settle in. But I expect to see more. I have long expected to catch glimpses of them as far as they went. Could you expect less from a boy who grew up in the woods and grew old as a schoolteacher and so spent most of his life staying close to the young who are elite and select and, by definition, often in trouble? I came to Mann Gulch expecting to catch glimpses of them as far as they could go. That's why I came.

The fire had gained on the crew at every stage of the race, until even the fastest had to fall. For a moment only the Four Horsemen were higher up the hill and still alive and for only a part of that moment would they see tragedy was among them. You can see tragedy coming from a considerable distance when you are older, but when you are young tragedy does not pertain to you and certainly never catches up to you. There are pieces of premonitions of tragedy floating around, but they do not yet add up to your tragedy. There are separate stabs of fear, of pity, of self-pity, but to a degree in separate parts of the body. Then suddenly they all merge into one sense, the encompassing sense of inevitability. It is everywhere on you as it becomes the essential whole of all that is preparing to be your tragedy. It becomes the cause of your mounting fear, your pity, and your self-pity, telling you that, no matter what, it does no good to be proud and good and young. Then, almost at the end, it makes possible the triumph that can come at the end of tragedy for the young who are select and elite—the triumph of retaining your pride when you know you have lost for good before you have had a chance in life to make good, except for this.

———

From the elevation of retrospect we can see it all coming together more clearly and sooner than those who were there and running. For us the signs are many that in minutes the blowup would bring a total convergence of sky, young men, and fire, and after that the dark; on the top of the hill, though, there were only occasional partings in the smoke, the flames themselves were blinding, and those inside the flames and smoke could no longer see what was happening to them and would happen next. We would not have started to follow the course of wildfire if we had not assumed that all of us, when called upon, could view an earthly scene from imaginative perspectives, something like the Sky Spirits in Thomas Hardy's poetical drama, *The Dynasts,* who comment upon tragedies of man from distant horizons. The title of another of Hardy's finest poems, "The Convergence of the Twain," suggests convergences yet to come in Mann Gulch. Hardy's convergence is between the elite, brightly lit, and fastest ship of its time, the *Titanic,* with an iceberg moving inexorably out of "a solitude of the sea / Deep from human vanity." We can set aside the difference between the ice of the poem and the fire of the story; from the beginning of the world either way has been taken as the way the world may end, and surely there is little philosophic difference between the convergences of fire and ice.

> Alien they seemed to be:
> No mortal eye could see

The intimate welding of their later history,

Or sign that they were bent
By paths coincident
On being anon twin halves of one august event,

Till the Spinner of the Years
Said "Now!"

There is, however, a mathematical difference between the perspective from a distant horizon and the view from the ground. From the ground, our approaching tragedy, like the *Titanic's*, had been linear, arithmetic, and two-dimensional. From the ground, it had occurred on one line as Behind caught up to Ahead, but the Spinner of the Years, viewing wildfire and young men from an even more distant horizon than our own, would see Geometry as well as Arithmetic in what was occurring at lower elevations. Not just Geometry but Solid Geometry—lines becoming curves and curves closing into circles and circles blowing up into spherical monsters whirling burning branches into the sky—and for a short time, a very short time, a thin line moving among these expanding solids.

At the appropriate moment when the Spinner of the Years said "Now," Sallee crawled through the crevice 170 yards above where he had left Dodge and took a good look back. He was the only survivor who stopped just to look. Although Sallee was not on a distant horizon and did not have the complete view of the Spinner of the Years, he saw three and possibly four geometries of fire: Below and Above him, Above being almost Beside him; and possibly Ahead of him in the head of the gulch already filling up with smoke.

And certainly Behind him were two fires, Mann Gulch's

two famous fires, the main one now only fifty yards or less downgulch and in front of it the escape fire. In fact, he saw all the preconditions of a tragedy only moments away, without realizing it was imminent and inevitable and tragic. Both he and Rumsey, after the fire had passed the rock slide where they had taken refuge, thought that the rest of the crew were probably as safe as they were. Maybe you have to be born with a special sense of the inevitable to see it coming.

On a big fire, fire could be everywhere, but you can't look everywhere or your problem gets unmanageable. Rumsey kept the problem simple by looking only for the top of the hill and nothing else, and he got there—only Sallee saw the lower edge of the main fire pass below the lower end of Dodge's escape fire, but evidently he thought it was a stray stream of fire and didn't let it bother him by thinking it might be a killer. The fire as it existed in his mind was something behind them that had jumped from the lower end of the opposite side of the gulch and was still behind them.

No matter who you are it is hard to adjust yourself to the fact that a forest fire is not all a big roar behind you getting closer—a dangerous part of it is very sneaky and may actually have sneaked ahead of you or is trying to and doesn't roar until it is about to close in on you. The fire that had jumped to the north side of the gulch had also been sneaking upgulch in the dense timber on the south side where it began, rolling burning cones downhill and setting spot fires, until in the semi-darkness something invisible touched something else invisible and suddenly there was a fire front surrounding the head of the gulch. In a few minutes the head of the gulch descended into the lower circles of its own Inferno and the blowup became complete in Mann Gulch.

It was about this time that ranger Jansson, looking up Mann Gulch from the river, saw that the upper end of the gulch had disappeared into one vast flame. By then the fire passing below the lower end of Dodge's fire, which looked like Behind to Sallee, had moved straight upgulch and merged with the fires that had become Ahead in a great flame, as Ahead was already disappearing in one flame out of the gulch. Below was soon to be Above as well as Behind and Ahead. It was not just a convergence of the twain but of a quatrain, for Above surely was rolling its torches downhill to meet the main fire spreading up from Below.

We are beyond where arithmetic can explain what was happening in the piece of nature that had been the head of Mann Gulch. Converging geometries had created something invisible like suction to carry off a natural explanation of the attraction of geometries to each other. In between these geometries for something like four minutes was a painfully moving line with pieces of it dropping out until there came an end to biology. Then it was pure geometry, and later still the solid geometry of concrete crosses.

There has been no final account as yet of the escape fire started by the foreman of the Forest Service. It is time now that there should be. Near the end of many tragedies it seems right that there should be moments when the story stops and looks back for something it left behind and finds it and finds it because of things it learned, as it were, by having lived through the story. The things found can be relatively small things, such as this thing, but also they can be big; but usually they are announced by minor characters, and generally they are about nature. We are so often wrong about nature that it comes as a relief of some kind to be right about it, especially after there has been some great disruption in it. Such moments of relief near the end of tragedy must be important parts of what from classical times has been called the purgation of tragedy. At times it seems as if tragedy tries at the end to take away some of its own tragedy, and if some tragedies never restore our stability, at least most of them allow us some success in struggling to attain some stability on our own. In my family, some such meaning was attached to the phrase "saved by grace." The remaining pages of this tragedy are its purgation and they come by grace. In my family, what happens on Sundays is foreordained. What comes on weekdays comes from something within us and for which we are responsible, and if it is from something deep within us it is called "grace," and is.

———

The Mann Gulch fire would never have attained its preeminence in the history of forest fires if foreman Wag Dodge had not set his escape fire. It made the Mann Gulch fire a lasting mystery story, unlike much larger tragic forest fires that were open-and-shut affairs, buried forever with only one interpretation. With only one interpretation, a forest fire soon becomes a statistic. But the Mann Gulch fire was not only a tragedy but a mystery story of physical and intellectual dimensions, introducing mistaken identities and explanations accompanied by seemingly unanswerable scientific questions.

The mistaken identities and explanations start with Dodge's fire. Sallee and Rumsey, the first to encounter it, mistakenly judged it to be a "buffer fire." To Dodge it was indeed intended to protect the crew from the main fire, but he intended that the men should lie facedown in its ashes with him and let the main fire burn over them. To those Dodge tried to persuade to lie down in the hot ashes with him, Dodge's fire seemed unintelligible. Strategically, all they could see was that he had started a fire even closer to them than the main fire. Who knows what they thought of him as a character? They may have thought he had panicked and gone nuts, or at least had lost his guts and turned chicken—and only two hundred yards from the top.

We have already encountered some of the bitterness aroused by the so-called escape fire in the hearts of those in the world outside who had been close to those who died by the crosses. As we know, the most bitter was Henry Thol, father of the jumper whose cross is closest to the top, who thought Dodge's setting a fire in front of the main fire was a homicidally incompetent act. It had killed his boy and other jumpers who were close to the top, and prevented others down the hill from escaping the only way that may still have been open to them.

It is only right for us to have Henry Thol, Senior, state again his case against the Forest Service and Dodge, reminding ourselves that, as a retired Forest Service ranger, he was the only first-class woodsman among the plaintiffs: "When [Dodge] set this fire, he didn't know what he was doing. Indications on the ground show quite plainly that his own fire caught up with some of the boys up there above him. His own fire prevented those below him from going to the top. The poor boys were caught—they had no escape."

Lurking in Thol's charges is a test that could have been applied to determine their validity but would have had to be applied immediately while the evidence was still fresh and discernible. Thol and Carl A. Gustafson, chief of the Division of Fire Control in Washington, both testified for opposite reasons at the Board of Review that it was easy to follow the outline of Dodge's fire on the hillside since it burned with much less intensity than the main fire. Powerful evidence, if it appeared as clear on the hill as it sounds in court. Why, then, didn't at least one of these leading exponents of opposite conclusions challenge the other to return immediately to Mann Gulch with a jury of experienced and impartial woodsmen to study the outlines of Dodge's fire before autumn rains and winter snow made mud of the evidence? Big talk, but no follow-up—maybe those outlines of Dodge's fire weren't as clear on the ground as the witnesses said they were.

There is another test, of course, that should allow the hillside to reveal the role Dodge's fire played in the tragedy. Gustafson, who was designated by the chief forester to conduct the first investigation of the Mann Gulch fire, and who before even seeing the gulch was worried about whether Dodge's fire had killed the men, almost immediately recognized the two questions to which right answers could clear the Forest Service of negligence. The two questions were directed again and again to Rumsey and Sallee: (1) Did you keep to the upgulch side of Dodge's fire on your race to the crevice after leaving Dodge? (2) Was the crevice where you passed to safety straight or almost straight above where you left Dodge? If Rumsey and Sallee were in front of Dodge's fire and still were able to run almost straight to the top of the ridge, Dodge's fire could not have caught the crew angling upgulch.

In their testimony Rumsey and Sallee more and more agreed that Dodge's fire went straight upslope, but they were young and were under pressure from officials of the Forest Service to agree to what ranger Jansson scornfully referred to as the "established" version of what had happened. Their increasing agreement needed the support of other evidence—and certainly of evidence on the ground. So almost from the beginning of our serious study of the Mann Gulch fire Laird Robinson and I realized that a positive location of the crevice and the site where Dodge and his crew had separated at his fire would be central to a determination of much of what had happened in the final acts of the tragedy, including the role of Dodge's fire. Properly locating the lower end of Dodge's fire not only was essential to reconstructing the story of the Mann Gulch fire but became something of a story in itself, a quest story, obstructed by staggering heat and rattlesnakes looking for holes in which to cool off. This quest story ended happily, with Laird and me, led on by a long forgotten photograph, coming to a place in dead grass where a wooden cross had fallen. What is more, the wooden cross pointed us to a new crevice in the reef above, and at the far end of the crevice to a juniper bush. From this crevice you can do what Sallee did—turn and look back and see that you climbed, not exactly straight up, but close enough to straight up to regard the location of the crevice and the site where Dodge lit his fire as evidence heavily in support of the testimony of the survivors that the fire set by Dodge did not burn his crew.

Indeed, the primary witnesses—Rumsey, Sallee, the wooden cross, and the crevice—seem so nearly in complete agreement that it is hard not to imagine them irritably demanding, "What more do you want? Why isn't our agreement conclusive proof?" But a major trouble remains—the agreement among witnesses seems to contradict nature. How can two large fires only fifty yards apart burn in two different directions, one almost directly across the front of the other? The question is made more difficult by the fact that one fire must have been at something approaching maximum speed and intensity, a blowup contributing to its own wind and moving up the gulch with enough speed steadily to gain on the crew, whereas the other fire had started just minutes before and could not have been burning with much intensity—yet this newer, weaker fire seemingly had in its weakness enough power to free itself from the forces that controlled the main fire and go off on its own, defiantly burning across the front of the main fire. Why, when the two fires were almost one, didn't the forces controlling the more powerful fire also control the weaker one?

This is one of those tough spots where either our facts are wrong or we don't know enough about the subject to explain them. Out of deference to our witnesses and ourselves and because we don't like to start out by admitting we are wrong, let us assume at first that there is nothing seriously wrong with what we "know," namely, the testimony and the ground, and that, therefore, to explain what we know, we have to put aside some of what we think we already know about nature (which does not contradict itself but just is, even if it blows up and goes in two different directions at once). Nature in this case is the action of two wildfires as they approach each other, one much stronger than the other.

More precisely, therefore, we may not know enough about the factors of nature determining the direction and speed of wildfires, chiefly fuels, the direction and velocity of the wind, and the slope or pitch of the ground. However, a good look at the ground in the general area where this phenomenon occurred reduces these three possibilities to one. There was no appreciable difference between the fuels the two fires were burning that would cause the fires to burn in different directions. For instance, Dodge's fire had no rock slide on its upgulch side that would have turned the fire straight upslope, no thick stand of timber, nothing like that; both fires were burning in dry, waist-high grass—for practical purposes, in the same grass. Likewise, there is no appreciable difference in the grade of the slope on which they were burning. Dodge's fire was not burning on a suddenly steep pitch on the hillside and the main fire had not come to a depression or plateau, the combined effect of which might have sent Dodge's fire much more upslope than the main fire. The slope they were both burning on was very steep and would certainly have had an upslope influence on both fires, but whereas that influence was the same on both fires it was not great enough to deter the main fire from moving quickly upgulch and soon passing over and out of the head of the gulch.

That leaves the wind as the leading suspect, although it doesn't come easy to picture the wind blowing the main fire one way and the new fire defiantly across its front as the two fires came together. That's the picture, though, we seem to be left with.

If the wind was responsible for such an action, it would be natural to think that the main, upgulch wind suddenly changed direction and burned upslope, then turned back and burned upgulch again. And this is what Sallee thought must have happened. In his statement to the Forest Service investigator he said, "There apparently was quite a high wind blowing straight up the slope at the time Dodge set the fire because that fire spread very rapidly straight up the slope and only slowly sideways of the slope." No doubt the state of knowledge about fire behavior at the time was such that nearly all firefighters familiar with the Mann Gulch fire would have had no other explanation to offer. But it is highly unlikely that a sudden shift in wind would completely control the direction of Dodge's fire and leave untouched the main fire only a few yards away, which in a few seconds would pass around the lower end of Dodge's fire to continue upgulch.

It is much more likely that the reason why Dodge's fire went straight for the top of the ridge is also one of the reasons, not generally understood at the time of the fire, that converging fires can explode into a blowup. In this modern explanation of the causes of these two natural phenomena

—a blowup and two fires going in different directions—the Mann Gulch fire with the passing of time comes to explain itself and helps to explain other things like it.

An explanation of the blowup that jumped the lower end of the gulch and pursued the crew upgulch depends upon an understanding not only of prevailing winds approaching a promontory but of the wind effect created when two bodies of air of unequal temperature approach each other. That effect may be more easily recalled if it is again called a "convection effect." A fire can set up a whirling action by drawing the cooler and heavier air from the outside into the vacuum left by its own hotter and lighter air constantly rising and escaping. In case this seems like a theoretical and theatrical construction, you might go to your basement furnace when it is roaring and open its door, put your face in front of it, and feel with sudden alarm that you are about to be drawn into your own furnace.

The problem, then, of the direction Dodge's fire took finally comes down to the question of which way or ways the wind blows when two fires approach each other, one much more intense than the other. Something like a big tug of war between two winds would have to take place to control Dodge's fire, the two winds being the prevailing wind that was driving the main fire upgulch, and an opposite, downgulch wind the main fire had generated by drawing to itself heavier, cooler air to fill the vacuum caused by its hotter, lighter air rising and escaping. If the two fires had been at a distance from each other, this secondary, downgulch wind would have had little effect on Dodge's fire, but as the main fire drew closer to Dodge and his men, its effect would have increased until a moment came when the upgulch and downgulch winds approached equal force. At that moment of relative equilibrium between opposing winds a moment

of relative calm would have fallen on the slope where Dodge
struck his match, and one match could burn long enough to
start a fire. You can safely bet that where and when Dodge lit
his fire with a single gofer match there was no wind blowing
thirty miles an hour.

In that moment of calm when the two counterwinds
neutralized each other and so roughly eliminated wind as
the major factor in determining the direction of Dodge's
fire, the hitherto lesser force of the steepness of the slope
took over, and Dodge's fire, now under the slope's influence,
burned straight or almost straight upslope. At that moment
the survivors, the ground, the fires, and the winds were all
in agreement—that is to say, testimony and nature were in
agreement, and nature while seeming to act unnaturally was
actually in agreement with all parts of itself.

But the moment of the equilibrium of the counterwinds
could not have lasted long. What then? The answer is that
the convergence of all these convergences in another few
minutes completed itself into total conflagration at the head
of Mann Gulch. Already the lower branch of the main fire
had swept around the lower end of Dodge's fire and, roaring
upgulch and upslope, was closing in on the crew from below.
Threatening the crew from above was the fire burning upgulch on top of the ridge. Now behind them were two other
fires about to converge, the main fire that was pursuing them
upgulch and the fire Dodge had lit that was now for the moment burning roughly at right angles to the direction of the
main fire, but soon would be taken over by the main fire and
be propelled with it toward the head of the gulch, one indistinguishable from the other. No wonder neither Gustafson
nor Thol offered to take the other or an impartial jury back to
Mann Gulch to find the lighter outline of Dodge's fire. With
fires by now jumping over or sweeping around each other,

or overpowering each other, there couldn't have been much left of outlines.

The fire burning along the top of the ridge may have kept pace with the angling crew below and forced them farther and farther upgulch. It may have been the front of Dodge's fire, turning toward the head of the gulch after reaching the top of the ridge. More likely, it was the upper branch of the main fire, or, as the convergence approached completion, both fires combined. When the survivors and Laird and I were in Mann Gulch in 1978, we split our four votes among these three alternatives, with each of us willing to admit he might be wrong and could never be sure that he was right. But in the conflagration that was about to occur, no component any longer had any individual responsibility for the simple reason that in a moment there were no individual components. Just conflagration. What was happening was passing beyond legality and morality and seemingly beyond the laws of nature, blown into a world where human values and seemingly natural laws no longer apply. Such moments can occur the world over, sometimes even at home as well as on hillsides.

The Mann Gulch fire was passing beyond issues and settlements into a world of pictures—perhaps more exactly into thoughts that pictorialize and feel and cannot reduce themselves to numbers. These are pictures made largely by us, the amateur artists who are always making pictures inside our heads (that spring from our hearts).

PART THREE

15

It would be natural near the end to try to divest the fire of any personal liability to those who died in it and to become for a moment a distant and detached spectator. It might be possible then, if ever, to see fire in something like total perspective as it became total conflagration. If you had known something about wildfire, you already would have seen spot fires twisting themselves into fire swirls and fire swirls converging upon themselves. But viewing total conflagration is literally blinding, as sight becomes sound and the roar of the fire goes out of the head of the gulch and away and beyond, far away. The last you saw of the ground was a mole coming out of the smoke, a little more terrified than you, debating which way to go and ending the argument with itself forever by turning back into the impenetrable fire. So it is, when you cannot see the fire because of the smoke, sight becomes sound. You hear the fire as a roar of an animal without the animal or as an attacking army blown up by the explosion of its own ammunition dump.

Pictures, then, of a big fire are pictures of many realities, designed so they change into each other and fit ultimately into a single picture of one monster becoming another monster. The pictures and the monsters are untroubled by mathematics. The monster becomes one as it extends itself simultaneously as a monster and a real animal or more likely just as a part of a real animal—after it disgorges itself, all that can be seen of it from afar are its fried gray intestines. Oddly, as destruction comes close to being total, destruction erects for brief moments into the sexual and quickly sinks back again into destruction. Intestines stretch out all the way to the curvatures of the brain. The two don't look much different, and they aren't and they are.

Thus, pictures which wildfire creates of itself are at least bi-visual, part of the fire's process of procreating its meanings. So, as the fire at the top of the ridge slithered through the rocks, it stretched itself out into a snake rearing its head to see that it was on course and using its tongue as a torch to cut through obstructions. So, too, a little lower on the hillside, when the main fire paused for a moment in front of the escape fire, the red flames crowded together until they became ravaging military monsters enraged by the precocity of an obstacle in front of them; then for a moment these deranged military monsters, blocked in their advance, raged sideways up and down the line looking for a way to pass— small fires were left behind as the phalanx of flame threw torches ahead, jumped the line, and left something like a smoldering monster in ruins.

Because of their many meanings, wildfires can be tri-visual or more. Some of what even a seasoned firefighter sees never seems real.

After its deranged military front had passed, pieces of the main fire remained burning fiercely in clutches of timber. Dead standing trees, especially Ponderosa pine full of resin, became giant candles burning for the dead. Then one would explode, disappearing from the air where it stood, detonated by its own heat. The disappearance of the tree would not be visible; it would be a theological disappearance; immediately after the explosion, its falling would be transubstantiated into spreading waves of earth generated by its own earthquake, and after its waves had swelled and broken and passed over and under and on, it would return as sound and terminate in echoes of its earthquake rumbling out of the sides and head of the gulch. The world then was more than ever theological, and the nuclear was never far off.

By now, if not sooner, the fire had become total—it was below, above, behind, and now also in front at the head of the gulch. Spot fires must have been burning there, started in the grass by burning cones and ashes blown from the approaching fires. Suddenly in the semi-gloom they would pop out of the ground, bi-visual as little poisonous mushrooms. The bi-visual mushrooms bred instantly, swelled with impregnation underground and aboveground into a vast bulbous head with a giant stalk. The vast wildfire continued on its bi-visual way—sometimes the bulbous mushroom looked like a bulbous mushroom impregnated by a snake in the grass and sometimes like gray brains boiling out of the crevice of the earth. Then the brains themselves became bi-visual and changed again into suffering gray intestines.

The atomic mushroom has become for our age the outer symbol of our inner fear of the explosive power of the universe. It is the symbol of a whole age, and it took an artist to express the meaning the mushroom has for us. Henry Moore, one of our age's most expressive sculptors, commemorated the occasion that led to the Atomic Age—the first self-sustained nuclear reaction—on the site at the University of Chicago where it occurred. His bronze atomic mushroom, with its hollow eyes, is intentionally bi-visual from every point of view. Wherever you stand, the bronze looks like both an atomic mushroom and a skull, and is meant to.

When the blowup rose out of Mann Gulch and its smoke merged with the jet stream, it looked much like an atomic explosion in Nevada on its cancerous way to Utah. When last seen, the tri-visual figure had stretched out and was on its way, far, far, far away, looking like death and looking back at its dead and looking forward to its dead yet to come. Perhaps it could see all of us.

No one could know the power of it. It stretched until it became particles on the horizon, where it may have joined the company of Sky Spirits as particles, knowing what we do not know, probably something nuclear.

Now, almost forty years later, small trees have just started to grow along the bottoms of dry finger gulches on the hillside in Mann Gulch, where moisture from rain and snow are retained underground. Since even now these little evergreens are only six or eight inches high, the grass has to be parted to find them, but I look for such things. I see better what happens in grass than on the horizon. Most of us do, and probably it is just as well, but what's found buried in grass doesn't tell us how to get out of the way.

At the end, our point of view of the fire changes radically so that we no longer look down from the distant horizon and see in the blinding smoke only pictures composed of our primitive history and our nuclear future. Instead, now at the end we stay as close to the ground as we can, are guided by our compassion, and accompany highly select young men who never once realized that they could be mortal on their way to the obliterating earth. We should hope, though, in trying to identify ourselves with them that we will be able to retain our own identity, for their sake as well as ours. Because we are much older than they were ever to be and have lived in a time more advanced scientifically than theirs, we should know much that they did not know but that should be of value in this journey of compassion. In a journey of compassion what we have ultimately as our guide is whatever understanding we may have gained along the way of ourselves and others, chiefly those close to us, so close to us that we have lived daily in their sufferings. From here on, then, in the blinding smoke it is no longer a "seeing world" but a "feeling world"—the pain of others and our compassion for them.

Things moved rapidly to the end after the crew left the foreman at his escape fire. It makes no difference whether the crew could not understand in the roaring of the main fire what Dodge was trying to say to them or whether they thought his idea of lying down in the hot ashes of his own fire was crazy. Either way they were entering No Man's Land, lonely in the boiling semi-darkness of the main fire, which by now must have been less than fifty yards behind them. Rumsey and Sallee, ahead of them, testified that the smoke parted enough for them to see the top of the ridge only two or three times. If we have difficulty seeing the rest of the men, they had difficulty seeing themselves. Heat and loneliness were becoming the only remaining properties. Their loneliness loomed up suddenly—they were young and not used to being alone, and as Smokejumpers they were not allowed to be alone, except in that perilous moment after they jumped from the sky and before they landed on earth.

It has been said since tragedy was first analyzed that it is governed by the emotions of fear and pity. As the Smoke-jumpers went up the hill after leaving Dodge it was like a great jump backwards into the sky—they were suddenly and totally without command and suddenly without structure and suddenly free to disintegrate and free finally to be afraid. The evidence is that they were not afraid before this moment, but now great fear suddenly possessed the empty places.

Beyond the world of sight and soon even beyond fear, the nonhuman elements of heat and toxic gases were becoming the only two elements, and soon heat was even burning out fear. To find a place that was cool was all that was left of human purpose. Knowing at least this much about fire and mortality, we can guess why most of the dead young men were found in depressions on the hillside; there it was thought to be cooler, so it was there that most of them went before they fell.

From our knowledge of others close to us we may learn something about how it felt this near the end. In the spring of the year my wife died from cancer of the esophagus, she remarked to me, "I feel as if I had spent all winter with my head under water." Later, when I asked a doctor what he thought it must be like to die in a fire, not from the burning but from the suffocation and lack of oxygen, he replied, "It is not terrible," and then added, but not as an afterthought, "It is something like drowning." If you compare my wife's remark to this more scientific attempt to speak of death by suffocation, you can see how careful my wife was, when she allowed herself to speak of such matters, to speak with precision.

It was not, therefore, for most of them, a terrible death. Many of the bodies were terribly burned when they were found, so much so that later it was discovered the caskets should not have been brought into the funeral chapel. Even so, they did not die of burning. The burning came afterwards.

To project ourselves into their final thoughts will require feelings about a special kind of death—the sudden death in fire of the young, elite, unfulfilled, and seemingly unconquerable. As the elite of young men, they felt more surely than most who are young that they were immortal. So if we are to feel with them, we must feel that we are set apart from the rest of the universe and safe from fires, all of which are expected to be put out by ten o'clock the morning after Smokejumpers are dropped on them. As to what they felt

about sudden death, we can start by feeling what the unfulfilled always feel about it, and, since the unfulfilled are many,
the Book of Common Prayer cries out for all of them and us
when it begs that we all be delivered from sudden death.

Good Lord, deliver us.

From lightning and tempest; from earthquake, fire, and
flood; from plague, pestilence, and famine; from battle and
murder, and from sudden death,

Good Lord, deliver us.

One thing is certain about these final thoughts—there
was not much size to them. Time and place did not permit even superior young men dying suddenly "to see their
whole lives pass in review," although books portray people
preparing to die as seeing a sort of documentary movie of
their lives. Everything, however, gets smaller on its way to
becoming eternal. It is also probable that the final thoughts
of elite young men dying suddenly were not seeing or scenic
thoughts but were cries or a single cry of passion, often of
self-compassion, justifiable if those who cry are justly proud.
The two living survivors of the Mann Gulch fire have told
me that, as they went up the last hillside, they remember
thinking only, "My God, how could you do this to me? I cannot be allowed to die so young and so close to the top." They
said they could remember hearing their voices saying this
out loud.

Of the two great tragic emotions this close to the end, fear
had been burned away and pity was in sole possession. Not
only is it heat that burns fear away; the end of tragedy puri-
fies itself of it. Before the end of a tragedy the most famous of

tragic heroes can stand in fear before ghosts and can shake in front of apparitions of those they have murdered, but by the end the same tragedy has purged these same tragic heroes of fear, as is made immortally clear by the last lines of one of the most famous of these tragic heroes: "Lay on, Macduff; / And damn'd be him that first cries, 'Hold, enough!'"

The pity that remains is perhaps the last and only emotion felt if it is the young and unfulfilled who suffer the tragedy. It is pity in the form of self-pity, but the compassion felt for themselves by the tragic young is self-pity transformed into some divine bewilderment, one of the few emotions in which the young and the universe are the only characters. Although divine bewilderment addresses its grief to the universe, it only cries out to it. It has to find its answer, if at all, in its own final act. It is not to be found among the answers God gave to Job in a whirlwind.

The most eloquent expression of this cry was made by a young man who came from the sky and returned to it and who, while on earth, knew he was alone and beyond all other men, and who, when he died, died on a hill: "About the ninth hour he cried with a loud voice, Eli, Eli, lama sabachthani?" ("My God, my God, why hast thou forsaken me?")

Although we can enter their last thoughts and feelings only by indirection, we are sure of the final act of many of them. Dr. Hawkins, the physician who went in with the rescue crew the night the men were burned, told me that, after the bodies had fallen, most of them had risen again, taken a few steps, and fallen again, this final time like pilgrims in prayer, facing the top of the hill, which on that slope is nearly east. Ranger Jansson, in charge of the rescue crew, independently made the same observation.

The evidence, then, is that at the very end beyond thought and beyond fear and beyond even self-compassion and divine bewilderment there remains some firm intention to continue doing forever and ever what we last hoped to do on earth. By this final act they had come about as close as body and spirit can to establishing a unity of themselves with earth, fire, and perhaps the sky.

This is as far as we are able to accompany them. When the fire struck their bodies, it blew their watches away. The two hands of a recovered watch had melted together at about four minutes to six. For them, that may be taken as the end of time.

It was 6:10 by Dodge's watch when he rose from the ashes of his own fire. From then on, Dodge had his own brief tragedy to lead, which in some ways also must be considered a part of this tragedy.

We leave the dead on the hillside with a promise made to me at the Office of Air Operations and Fire Management of Region One of the United States Forest Service that their crosses will always be renewed.

I, an old man, have written this fire report. Among other things, it was important to me, as an exercise for old age, to enlarge my knowledge and spirit so I could accompany young men whose lives I might have lived on their way to death. I have climbed where they climbed, and in my time I have fought fire and inquired into its nature. In addition, I have lived to get a better understanding of myself and those close to me, many of them now dead. Perhaps it is not odd, at the end of this tragedy where nothing much was left of the elite who came from the sky but courage struggling for oxygen, that I have often found myself thinking of my wife on her brave and lonely way to death.